How and Why Species Multiply

Princeton Series in Evolutionary Biology
Series Editor: H. Allen Orr

Evolution lies at the heart of the life sciences. In the Princeton Series in Evolutionary Biology, outstanding scientists tackle central problems in evolution through monographs, texts, and synthetic studies. Titles will be drawn from the following areas: evolutionary genetics and genomics; molecular population genetics; theoretical evolutionary biology; computational and statistical genetics and genomics; molecular evolution; experimental evolution; phylogenetics and phylogeography; paleobiology; and macroevolution.

Dynamics of Cancer: Incidence, Inheritance, and Evolution
Steven A. Frank

How and Why Species Multiply: The Radiation of Darwin's Finches
Peter R. Grant and B. Rosemary Grant

How and Why Species Multiply

The Radiation of Darwin's Finches

PETER R. GRANT AND B. ROSEMARY GRANT

PRINCETON UNIVERSITY PRESS PRINCETON AND OXFORD

Library of Congress Cataloging-in-Publication Data

Grant, Peter R., 1936-
How and why species multiply : the radiation of Darwin's finches /
Peter R. Grant and B. Rosemary Grant.
p. cm.
Includes bibliographical references and index.
ISBN 978-0-691-13360-7 (hardcover : alk. paper)   1. Finches—Evolution—Galápagos
Islands.   2. Finches—Adaptation—Galápagos Islands.   I. Grant, B. Rosemary.   II. Title.
QL696.P246G733 2008
598.8′8—dc22   2007005384

British Library Cataloging-in-Publication Data is available

This book has been composed in Adobe Caslon

Printed on acid-free paper.
press.princeton.edu

Printed in the United States of America
3   5   7   9   10   8   6   4   2

Contents

CONTENTS

Illustrations

Color Plates (following page 122)

Tables

Preface

This book is dedicated to the memory of David Lack, a pioneer of a field that has come to be known as Evolutionary Ecology. Sixty years ago he published his first book, entitled *Darwin's Finches*. Based on four months of fieldwork in the Galápagos and many hours of measuring specimens in museums, it attempted to explain the ecology and evolution of an adaptive radiation made famous by Charles Darwin. It showed how it was possible to use observations of living animals to infer and interpret their evolutionary history. In so doing it established a new field of enquiry.

David Lack died in March of 1973. In a sense we feel we are the bearers of a torch he passed on, as our field research on the same birds began one month earlier. We had discussed Darwin's finches with him in Oxford two years before, but at that time we had no intention of studying them ourselves, and when we started we had no intention of continuing for 34 years.

Part of the success of his book was due to its simple prose, succinctness, and relative freedom from jargon. We set out to write the kind of book he would have written if he had had our experience, a book about the evolution of Darwin's finches in 150–200 pages. Our goal, like Lack's, was to capture the essentials and the highlights for an intended audience of students. To achieve this goal we have had to reduce to a minimum the historical background, and methods of study, analysis, and statistics; these can be found in the numerous cited papers and two books on Darwin's finches (Grant and Grant 1989, Grant 1999). In writing this book we also had in mind readers from a variety of backgrounds with a perennial sense of wonder and curiosity in the amazing diversity of birds and other organisms. Therefore, in order to make the book accessible to a broad audience, we have provided a glossary of the technical terms we could not avoid using.

Like everyone else, David Lack was influenced by his predecessors, especially by his mentor Julian Huxley (Huxley 1940, 1942) and by Robert Perkins (1903, 1913). Perkins deserves special mention because, like Lack, he went to an archipelago without prior training at doctoral level and used his eyes to make sense of an adaptive radiation; in his case it was the honeycreepers of

Hawaii. The ideas he developed on allopatric speciation, competition for resources, and niche specialization in food-limited birds, in contrast to the lack of specialization in enemy-limited beetles, must have influenced David Lack. These were not always acknowledged explicitly in Lack's book (Grant 2000a). There is a parallel to this with Charles Darwin, whose *Origin of Species* may have been more influenced by his grandfather Erasmus Darwin and by his mentor Robert Grant than he acknowledged. We may be unwittingly doing the same with our own sources.

We have reached many of the same conclusions as David Lack. Almost inevitably there are some differences. Four are worth mentioning here.

First, he believed that fossils were necessary to establish the patterns of evolution in the past. In the preface to the reissue of his book in 1961 he wrote: "Without fossil material to give perspective in time, it is impossible to be sure of which species are near the base and which near the top of the evolutionary tree." The revolution in molecular biology since 1961 has shown how it is possible to estimate the position of each species in an adaptive radiation.

Second, Lack believed the problem to be explained with Darwin's finches was the persistence rather than the origin of species. He adopted the prevailing view (Stresemann 1936, Dobzhansky 1937, Huxley 1938, Mayr 1942) that speciation was inevitable with the lapse of time; given geographical isolation that promoted diversification, all that was required was enough time for sterility to evolve and so, at secondary contact, interbreeding would be zero or close to it. Therefore speciation would be completed in allopatry, i.e., in geographical isolation on separate islands. However, when species so formed encountered each other some time later they would compete for food, making it unlikely they would coexist, hence the problem of explaining why species persisted. The problem is essentially ecological (Schluter 2000). His proposed solution, like Perkins's (1903), was evolutionary divergence in beak size and a concomitant reduction in competition. This solution has been upheld by modern research, but one of the premises of his argument is now known to be wrong. Sterility factors have not evolved and interbreeding does occur.

Third, he sought evidence for hybridization, failed to find it, and concluded it was not an important factor in explaining the large morphological variation in many of the finch populations or their evolution. Observations on the breeding of banded birds were needed to show (in 1976) that this was wrong (Grant and Price 1981, Boag and Grant 1984a), and indeed we now think that hybridization, far from being negligible, has been a potent force in the adaptive radiation.

Fourth, he considered songs to be unimportant as signals of species identity because they are not discretely different. Modern work, starting with Robert Bowman (1979, 1983), shows they are important, even though individuals of one species occasionally acquire the song of another.

Lack cannot be faulted for not knowing as much as we do. He did not have computers, gels for electrophoresis, tape recorders, or the opportunity to detect rare hybridization that comes from a long-term study. Yet his ecological insights have been substantiated so often that one student of Darwin's finches (Schluter 2000) has suggested the birds should really be called "Lack's finches"!

We acknowledge a huge debt to the many people who have helped us in research over many years. The help began with advice from Robert Bowman and a four-month season of field research by Ian and Lynette Abbott, followed by several field seasons of research by Laurene Ratcliffe, Peter Boag, Trevor Price, Dolph Schluter, Stephen Millington, Lisle Gibbs, Lukas Keller, and Ken Petren, as well as many assistants. Our daughters Nicola and Thalia have helped in numerous ways; in Thalia's case for as long as we have worked on the Galápagos. We were initially funded by McGill University, and then on a continuing basis by the National Science and Environmental Research Council of Canada and the National Science Foundation of the United States while we held positions at McGill University, University of Michigan, and Princeton University. The research would not have been possible without the continued support of the Charles Darwin Foundation, the Charles Darwin Research Station, and the Galápagos National Parks staff. The book manuscript was read in various parts by Margarita Ramos and Dolph Schluter, and throughout by three reviewers and Sam Elworthy. We are grateful for their numerous corrections and helpful suggestions. We thank Dimitri Karetnikov for invaluable help with illustrations.

January 2007

How and Why Species Multiply

The Biodiversity Problem and Darwin's Finches

Now it is a well-known principle of zoological evolution that an
isolated region, if large and sufficiently varied in topography, soil,
climate and vegetation, will give rise to a diversified fauna according
to the *law of adaptive radiation* from primitive and central types.
Branches will spring off in all directions to take advantage of every
possible opportunity of securing foods.
(Osborn 1900, p. 563)

I have stated that in the thirteen species of ground-finches, a nearly
perfect gradation may be traced, from a beak extraordinarily thick, to
one so fine, that it may be compared to that of a warbler.
(Darwin 1839, p. 475)

BIODIVERSITY

WE LIVE IN A WORLD so rich in species we do not know how many
there are. Adding up every one we know, from influenza viruses
to elephants, we reach a total of a million and a half (Wilson
1992, ch. 8). The real number is almost certainly at least five million, perhaps
ten or even twenty, and although very large it is a small fraction of those that
have ever existed; the vast majority has become extinct. Knowledge of the
world's biological wealth is constantly expanding—for example, new species of
marine fish are found each week. Nevertheless the rate of discovery of fundamentally different organisms is slowing down, and the discovery of a new
order, class, or phylum is an extremely rare event. New findings are incorpo-
rated into an existing Linnean framework, and rarely change it. In short, incomplete as the biological inventory of life on earth is, enough is known to

pose a strong challenge to evolutionary biologists. Explain it! Why are species as diverse as they are, and why are there so many?

The challenge is currently being met in numerous and diverse studies, for example in studies of butterfly wing pattern (Jiggins et al. 2006), plant phylogenies (Soltis et al. 2005), and fossil whales (Gingerich 2003), to name just three. Our own way of addressing the challenge is to start low down in the Linnean hierarchy, at the level of populations, species, and genera, reasoning that they contain the seeds of differences at the higher levels of classes, phyla, and kingdoms. The lessons learned at lower levels can then be extrapolated to higher levels where evolutionary pathways connecting related taxa are less clear. We choose a single group of related species for close scrutiny, and attempt to answer the following questions: where did they come from, how did they diversify, what caused them to diversify as much as they did (and no more), and over what period of time did this happen?

THE CHOICE OF ORGANISMS

Ideally, as well as for convenience, the group should be more than a few but less than a multitude. Preferably they should live in the same geographical location in which they evolved as this helps us to interpret their past evolution. They should be easy to study in captivity and in nature, and they should have left a good, recoverable, and interpretable fossil record of their history.

Organisms that come closest to meeting all these needs are members of adaptive radiations. An adaptive radiation is the rapid evolution from a common ancestor of several species that occupy different ecological niches (Givnish and Sytsma 1997, Schluter 2000). The organisms are numerous enough for quantitative comparisons, similar enough to enable us to reconstruct their routes of diversification, and they live in environments where those routes can be interpreted adaptively (or otherwise).

Prime candidates for study are species-rich genera that live in the same region. There are many of these. Some of the best known and impressively diverse are the cichlid fish of African Great Lakes (Kocher 2004, Joyce et al. 2005, Seehausen 2006), *Anolis* lizards (Losos 1998) and *Eleutherodactylus* frogs (Hedges 1989) of the Caribbean and Central and South America, and *Drosophila* (DeSalle 1995) and the Silversword alliance of Composite plant species (Barrier et al. 1999) of the Hawaiian archipelago. Several groups comprise hundreds of species, literally: more than 700 in the single genus of *Eleutherodactylus* alone (Crawford and Smith 2005), and the total is closer to a thousand in the case of Hawaiian *Drosophila* (Kaneshiro et al. 1995, Kambysellis and Craddock 1997). Then there are numerous Central and South American species of butterflies in the genus *Heliconius* (Mallet et al. 1998), *Partula* snails in Polynesia (Johnson et al. 2000), dipterocarp trees in Asia (Ashton 1982), figs and fig wasps distributed widely in the tropics (Weiblen 2002), likewise orchids and orchid bees (Pemberton and Wheeler 2006), etc., etc. The list goes on and on, and whether the groups of species meet strict or relaxed criteria for being recognized as adaptive radiations (Schluter 2000) they are certainly rich in species and diverse.

A more compact and manageable group than all of these is a small number of remarkable birds known as Darwin's finches (Plates 1 and 2). They are unique in what they offer biologists. They are so similar to each other that transformation of one species into another can be reconstructed easily. They are accessible; their behavior can be studied easily because they are tame. And importantly, no species has become extinct through human activities.

DARWIN'S FINCHES

Apart from a single species on Cocos island, Darwin's finches (subfamily Geospizinae) are confined to the Galápagos islands of Ecuador (Fig. 1.1). Depending on how they are classified there are 14 or 15 of them (Table 1.1), which is a convenient number for complete study. They constitute a classical case of adaptive radiation (Fig. 1.2), having been derived from a common ancestor, and diversified relatively rapidly in morphology and ecology. They live in the same, largely undisturbed, environment in which they evolved, consequently whatever we can learn about their ecology and evolution gives us insights into the process of speciation and adaptive radiation under entirely natural conditions. For instance, populations of the same species occur on different islands (Table 1.2), and in some cases they have different ecologies. This enables us to investigate the reasons for their divergence. Then again, closely related species occur together on the same island, and differ. This allows us to investigate the nature of the reproductive barrier between them, and the question of how and why species stay apart. So, considering populations across the entire archipelago, it is as though the whole process of

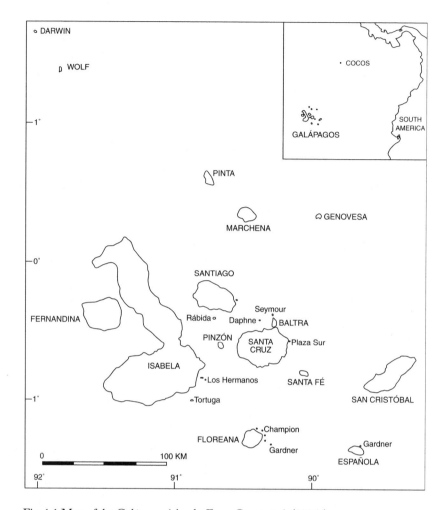

Fig. 1.1 Map of the Galápagos islands. From Grant et al. (2005a).

speciation is represented in all of its stages from start to finish: from an initial divergence to reproductive isolation, repeated many times.

These are very large advantages for their study. They are offset to some extent by two disadvantages: by the limited degree to which finches can be used experimentally, and by the absence of fossils except for very recent ones. As we shall see, some experimental investigations are possible, and molecules can sometimes help where fossils fail.

TABLE 1.1

Darwin's finch species. *C. olivacea* could be considered two species on the basis of genetic evidence (Fig. 2.1).
The woodpecker and mangrove finches have been considered members of a separate, weakly differentiated genus *Cactospiza* (Grant 1999).

Scientific Name	English Name	Approximate Weight (grams)
Geospiza fuliginosa	Small Ground Finch	14
Geospiza fortis	Medium Ground Finch	20
Geospiza magnirostris	Large Ground Finch	34
Geospiza difficilis	Sharp-beaked Ground Finch	20
Geospiza scandens	Cactus Finch	21
Geospiza conirostris	Large Cactus Finch	28
Camarhynchus parvulus	Small Tree Finch	13
Camarhynchus pauper	Medium Tree Finch	16
Camarhynchus psittacula	Large Tree Finch	18
Camarhynchus pallidus	Woodpecker Finch	20
Camarhynchus heliobates	Mangrove Finch	18
Platyspiza crassirostris	Vegetarian Finch	35
Certhidea olivacea	Warbler Finch	8
Pinaroloxias inornata	Cocos Finch	16

DIVERSITY OF DARWIN'S FINCH SPECIES

As a starting point for a discussion of their evolution we have followed the classification of species developed by Lack (1945, 1947) on the basis of an exhaustive study of museum specimens. Samples of specimens from the same island tend to fall into discrete groups: different species. Differences between species persist, even if they interbreed rarely, which is the essence of the biological species concept (Wright 1940, Mayr 1942).

The species differ in plumage and morphology (Plate 2). Those differences are summarized as follows. Six species of ground finches (genus *Geospiza*) are alike in having brown and streaked female plumage and unstreaked black male

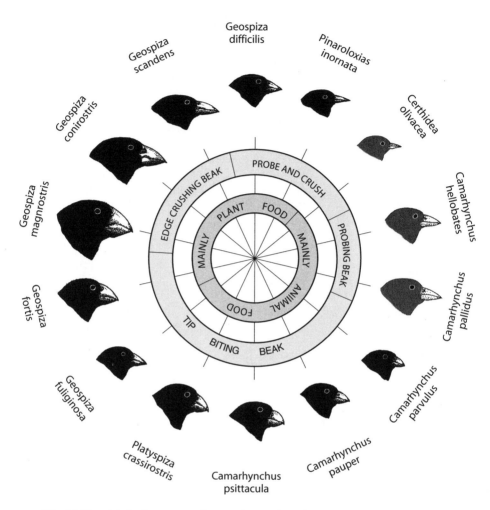

Fig. 1.2 Darwin's finches portrayed to emphasize the radiation, and not their genealogy (see Plate 1), like the canopy of a tree viewed from above. From Grant (1999).

plumage. Plumage traits set them apart from a group of five species of tree finches (*Camarhynchus*), which have a more olive-green than brown tone and few or no streaks. Males of three of them have black heads, shoulders, and chest, whereas males of the other two (*C. pallidus* and *C. heliobates*) have no black color, and nor do warbler finches (*Certhidea*). The vegetarian finch (*Platyspiza*) combines the color of ground finches with the limited expression

TABLE 1.2

Finch distributions on the 18 major islands.

B = breeding, (B) = probably breeding, E = extinct, (E) = probably present as a breeding population formerly, now extinct. The *olivacea* and *fusca* lineages of *Certhidea* are shown separately.

	G. magni-rostris	G. fortis	G. fuligi-nosa	G. difficilis	G. scandens	G. conir-ostris	C. psitt-acula	C. pauper	C. par-vulus	C. pallidus	C. helio-bates	P. crassi-rostris	C. olivacea	C. fusca
Seymour		B	B		B									
Baltra		B	B		B									
Isabela	B	B	B	(E)	B		B		B	B	B	B	B	
Fernandina	B	B	B	B			B		B	(B)	E	B	B	
Santiago	B	B	B	B	B		B		B	B		B	B	
Rábida	B	B	B		B		(E)		(E)			(E)	B	
Pinzón	B	B	B		(B)		E		B	B		E	B	
Santa Cruz	B	B	B	E	B		B		B	B		B	B	
Daphne	B	B	B		B									
Santa Fe	B	B	B		B		B		B			(E)		B
S.Cristóbal	E	B	B	(E)	B				B	B		B		B
Española			B			B								B
Floreana	E	B	B	E	B		B	B	B			B		E
Genovesa	B			B		B								B
Marchena	B	B	B		B		B					B		B
Pinta	B	B	B	B	B		B					B		B
Darwin	B			B										B
Wolf	B			B										B

of black of the male in three of the *Camarhynchus* tree finches. Finally the single species on Cocos (*Pinaroloxias*) shares the plumage features of the Galápagos ground finches.

Species in each of the ground finch and tree finch groups differ from each other to some extent in body size, but more so in beak size and shape (Fig. 1.2). That is how we can tell them apart: by their appearance. Furthermore the warbler, vegetarian, and Cocos finches possess beak morphologies quite unlike any of the others. As a result no two species in the entire group of Darwin's finches have the same beak morphology.

These few remarks capture the essence of three axes of variation in the radiation of Darwin's finches; a major beak axis, and lesser axes of body size and plumage. The variation is continuous in the case of beak and body size variation, and discrete and clustered in the case of plumage. An important feature is the placement of the species on the axes. Some species are surprisingly close together, such as the small, medium, and large ground finches (Plate 2). Others are more distant, and the warbler finch is distant from all others. Considered as a whole, Darwin's finches vary from a warbler-like bird of 8 g to a grosbeak-like bird of 35 g.

Their feeding ecology matches their morphological diversity, which is why the radiation of Darwin's finches is described as adaptive (Lack 1947, Bowman 1961). The ground finches feed often on the ground, consuming a variety of seeds, arthropods, as well as the fruits and seeds of prickly pear (*Opuntia*) cacti. Tree finches are more arboreal and insectivorous. The vegetarian finch is well-named for its vegetarian diet, and warbler finches feed on nectar and a variety of spiders as well as insects of small size.

Species and Populations

The simple morphological descriptions above belie complexity in the assignment of some populations to species. Although species are discrete when co-occurring on the same island, the boundaries between species are not always clear-cut when considered across the archipelago. This is to be expected in young, ongoing, adaptive radiations, and more will be made of it in later chapters. Blurring of the discrete morphological boundaries is what makes the finches such a promising but challenging group for the study of speciation.

Darwin (1842) wrote about this seven years after he, FitzRoy (Captain of the *Beagle*), and their assistants collected the first specimens in 1835:

> The most curious fact is the perfect gradation in the size of the beaks of the different species of Geospiza.

which led him to a prophetic evolutionary conjecture:

> Seeing this gradation and diversity of structure in one small, intimately related group of birds, one might fancy that, from an original paucity of birds in this archipelago, one species has been taken and modified for different ends.

Translated into the language of evolutionary biology, "modified for different ends" means adaptation by natural selection.

Complexity in the assignment of some populations to species occurs in two ways. First, populations of the same species differ from one island to another (Fig. 1.3). Although species are discretely different on the same island, as illustrated with ground finch species on Marchena in Figure 1.3, a large member of a small species on one island can be very similar to a small individual of a larger species on another island (Lack 1945, 1947, Grant et al. 1985). For example, medium ground finches on Santa Cruz are larger on average and vary much more than elsewhere. As a result the largest individuals of this population have larger beaks than the smallest of the large ground finches (*G. magnirostris*) on Rábida (Fig. 1.3). Thus, where to draw the line between the species is not always easy to decide. Nevertheless, despite some confusing similarities, individuals on an island are rarely difficult to classify (Lack 1947).

Second, well-differentiated populations on two islands could be considered separate species. There is no clear, unambiguous way of deciding whether members of such populations would interbreed and therefore whether the populations merit recognition as one species (conspecific) or two (heterospecific). The two prime examples are in the ground finch group. Six populations of sharp-beaked gound finches (*G. difficilis*) are united by shared features of beak shape but differ from each other in size enough to raise that question. For example, on Genovesa the average weight is 12 g, whereas on Santiago it is 27g! Populations of the large cactus finch (*G. conirostris*) on Genovesa and Española are morphologically similar to each other more than either is to the cactus finch (*G. scandens*), but they do

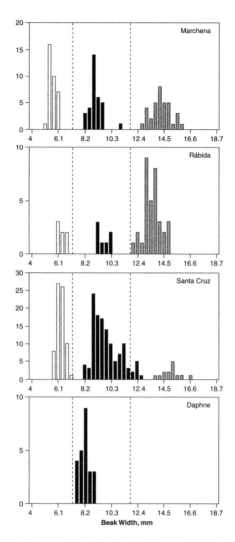

Fig. 1.3 Frequency distributions of beak widths of three species of ground finches, *G. fuliginosa* (white), *G. fortis* (black), and *G. magnirostris* (grey). They are approximately symmetrical and widely spaced apart on most islands (e.g., Marchena and Rábida), but not on Santa Cruz. Here *G. fortis* is exceptionally variable, and the largest individuals are larger than the smallest *G. magnirostris* on Rábida. The possibility of hybridization of *G. fortis* and the much rarer *G. magnirostris* on Santa Cruz is suggested by the shape of the *G. fortis* distribution, which is strongly skewed towards *G. magnirostris*. On Daphne Major the distribution of *G. fortis* beak sizes is shifted in the oppposite direction, in the virtual absence of the other two species (chs. 5 and 6). Based on measurements of male specimens in museums (Grant et al. 1985).

not coexist with the cactus finch anywhere, so possibly they should be considered conspecific with it.

The first of these complications raises questions about the identity of individuals, the second raises questions about the identity of species. They are not just classificatory conundrums. They are manifestations of the core Darwinian problem of understanding how species form. Paraphrasing Dobzhansky (1937) and Mayr (1942), that problem is how one population of interbreeding individuals splits into two with little or no interbreeding between them: how they multiply and diversify, and why.

Overview of the Book

In this book we attempt to explain the evolutionary diversification of Darwin's finches in terms of geography, behavior, ecology, and genetics. The explanation involves natural and sexual selection, random genetic drift, exchange of genes through hybridization (introgression), and cultural as well as genetic evolution. Linking all these factors together is the frequent and strong fluctuation in climatic conditions, with droughts on the one hand and extremely wet (El Niño) conditions on the other. An important conclusion will be that environmental change is an observable major driving force in the origin of new species. Using information from the study of contemporary finches, we then turn to questions of how and why the radiation unfolded in the way that it did. We focus on how environmental change in the past has guided the multiplication of finch species, and how some properties of the finches may have predisposed them to diversify.

We start (chapter 2) by using molecular genetic data to estimate phylogenetic relationships, that is the genealogical relationships between populations and species of Darwin's finches, and between them and possible relatives in Central and South America. Changes in the environment that have occurred since the finches arrived on Galápagos are described. This information sets the scene for considering how speciation could have taken place in theory (chapter 3). Subsequent chapters explore the steps with data. First we use ecological information to describe what happens when an island is colonized and a new population becomes established (chapter 4), how adaptation through natural selection occurs (chapter 5), and how competition for resources contributes to natural selection and evolution (chapter 6). Then we address the important

11

question of how individuals choose mates and what constitutes the barrier to interbreeding between species (chapter 7). Sometimes the barrier is breached, and interbreeding ensues. We examine the causes and consequences of hybridization in chapter 8. Chapter 9 uses the information and ideas developed so far to confront the issue of how species should be recognized. Chapter 10 focuses on the differences between species formed early and late in the radiation, and attempts to explain them in terms of responses to changing ecological opportunities and the balance between speciation and extinction. In chapter 11 we offer some explanation for why Darwin's finches radiated when other birds in the same environment did not. We place the radiation of Darwin's finches in a wider context in chapter 12 by outlining three stages that radiations pass through. Darwin's finches exemplify the first stage. We conclude this chapter with a short synthesis: a synthetic theory of adaptive radiation. Chapter 13 summarizes the main features of the Darwin's finch radiation, highlights what we do not know, and suggests some directions for future study.

Origins and History

By using knowledge from the past we may be able to accelerate
research into the mechanisms of adaptation to changing environments.
(Davis et al. 2005, p. 1713)

It is the intertwined and interacting mechanisms of evolution and
ecology, each of which is at the same time a product and a process,
that are responsible for life as we see it, and as it has been.
(Valentine 1973, p. 58)

INTRODUCTION

THE PRESENT IS KNOWN, the past is inferred. Evolutionary study of
biological diversity typically starts at the end, and works back to the
beginning: from the product of diversification to its origin, from verifi-
able observation to historical inference and implication.

Fossils of Darwin's finches date back a few thousand years at most (Steadman
1986), and are therefore not useful for answering questions about the origin of
the finches. In the absence of older fossils our best sources of information
about finch history are their genes. In this chapter we use genetic relatedness
among contemporary species to estimate their phylogenetic relationships and
history. The information is organized around three basic questions that need
to be answered to gain an understanding of the evolutionary history of this or
any other group of organisms (Grant 2001). The first question is "when
did the radiation begin," because the answer sets the boundary to relevant
Galápagos history. The second question is "what is the temporal pattern of the
radiation," and the pattern of affinities among the species, because the answer
reveals the particular evolutionary transitions that need to be explained. The

third question is how has the environment changed as the radiation proceeded, because a history of the ecological niches is needed to interpret the radiation. Paraphrasing Evelyn Hutchinson's felicitous language (Hutchinson 1965), we need to know how the evolutionary play unfolded as the scenery of the Galápagos theater changed. The answer to the first question is that the radiation is young, on the order of a couple of million years old, yet the environment in which it occurred has undergone large changes in the number of islands, climate, and vegetation.

PHYLOGENY

We have used mitochondrial and nuclear (microsatellite) DNA markers to estimate genetic relatedness among all Darwin's finch species, and between them and a group of related species on the South American continent and in the Caribbean. Measures of degrees of genetic relatedness have then been used to construct a phylogenetic tree (Fig. 2.1; for details of how this is typically done, see Schluter (2000) and Price (2007)). Some questions about the evolutionary history of the finches can be answered clearly from the tree, but others cannot because the species are genetically very similar to each other (Freeland and Boag 1999a, 1999b, Petren et al. 1999, 2005, Sato et al. 1999) and some of the relationships are statistically doubtful. Estimation of phylogeny with a mitochondrial DNA marker yields similar relationships to those shown in the figure, but with one important difference: the Cocos finch branches off later, from the base of the tree finch lineage (Petren et al. 2005). Nonetheless most of the species that group together on the basis of their genetic similarity with either microsatellite or mitochondrial markers match the groupings established in the classical era of taxonomy based on morphology (Lack 1947), and on allozyme (enzyme) similarities (Yang and Patton 1981, Stern and Grant 1996). A simplified representation of the relationships is illustrated in Plate 1.

One question about the origin of Darwin's finches can be answered unambiguously. All analyses have found Darwin's finches, including the Cocos finch, to be more closely related to each other than to any species elsewhere. Therefore they are monophyletic: that is, they all share the common ancestor that colonized Galápagos, and share it with no other species.

14

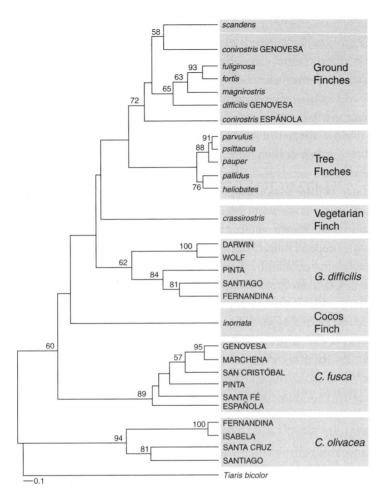

Fig. 2.1 Evolutionary relationships among populations and species of Darwin's finches reconstructed with microsatellite DNA markers. The numbers indicate statistical support for the nodes based on repeated sampling (bootstrap values). Species names (from Table 1.1) are in italics, island names are in capitals. Notice that the two populations of *G. conirostris* do not cluster together, nor do all populations of *G. difficilis*: hybridization (ch. 8), combined with the lack of time for strong genetic differences to arise, may be responsible (chs. 8 and 9). The root of the tree is provided by *Tiaris bicolor*. Adapted from Petren et al. (2005).

ANCESTORS

The identity of the common ancestor is uncertain. Darwin's finches are most closely related to a group of seed-eating tanagers in the Caribbean and Central and South America (Petren et al. 1999, Sato et al. 1999, 2001, Burns et al. 2002). Of the several candidates that have been examined so far no single species jumps out as the closest existing relative of modern Darwin's finches. There are half a dozen possibilities in the genera *Tiaris, Melanospiza* (Plate 3), and *Loxigilla* (Burns et al. 2002). Measures of genetic similarity show them to cluster together as a group. When current geographical proximity is combined with genetic similarity, members of the genus *Tiaris* have the strongest claim to be genealogically closest to Darwin's finches. The dull-colored grassquit *Tiaris obscura* could even be the ancestor itself (Sato et al. 2001).

The oldest of the Darwin's finches, the warbler finch (Plate 4), has a small and slender beak unlike any of the seed-eating relatives on the continent or in the Caribbean. If the ancestral species was similar to the warbler finch it is now extinct on the Caribbean and mainland. Alternatively, if the original species was like *Tiaris obscura* (Sato et al. 2001) or a relative (Burns et al. 2002) it gave rise to the ecologically specialized warbler finch, and became extinct on Galápagos. This seems more probable to us. Moreover, there are additional reasons for supposing that finch species have become extinct there, as we discuss in chapter 10.

THE TIME OF ARRIVAL

According to microsatellite DNA data the Cocos finch branched off from the phylogeny after the first division into two groups of warbler finches. The mtDNA data suggest a later origin (Petren et al. 2005). Either way the Cocos finch was derived from birds on Galápagos, and did not give rise to them. Therefore the Galápagos archipelago was colonized first, and Cocos island was colonized from Galápagos. We cannot know when the initial colonization happened, but we can estimate the time when the first split in the radiation took place, in other words when the radiation started.

Warbler finches occupy the basal position of the phylogeny, and they split into two groups (Plate 4) before they gave rise to any others. This is estimated to have happened 1.6–2.0 million years ago (MYA) (Petren et al. 2005). To arrive at this estimate we have assumed that the genetic difference between any two populations or species is proportional to the time since they became separated from each other and began to diverge; on average their rate of divergence is approximately constant. If, following current practice (e.g., Garcia-Moreno 2004, Lovette 2004), we assume that nucleotide sequences in the cytochrome b gene of mitochondrial DNA diverge in separate lineages solely through random changes in clock-like fashion, at an average rate of 2% sequence divergence per million years, then the first divergence in the Darwin's finch phylogeny occurred 1.6–2.0 million years ago (MYA). The first divergence was earlier, 2.1–2.5 MYA, if instead of 2% we use the well-calibrated rate of 1.6% sequence divergence per million years estimated by Fleischer and McIntosh (2001) for Hawaiian honeycreeper finches.

There are reasons for thinking this might underestimate the age of the first split. For one thing, allozyme differences indicate an older age ($\geq$ 2.8 MY) (Grant 1999), and for another, hybridization and gene exchange (ch. 8) after the initial divergence of two lineages reduces their genetic difference and hence makes the separation appear more recent than it really was.

Before 2 million years ago Darwin's finches existed on the islands for an unknown length of time as a single species. If *Tiaris* is the closest relative (Petren et al. 1999, Sato et al. 1999, 2001), Darwin's finches began their independent evolution about 2.3 MYA. However, the mitochondrial clock ticks at different rates in different groups of birds, and estimating the rate is especially uncertain for recently evolved species (Ho et al. 2005) such as Darwin's finches. Caution is needed for the additional reason that divergence in mitochondrial DNA between species can be halted and even reversed if introgressive hybridization occurs and the newly introduced mtDNA is selectively favored (Bachtrog et al. 2006). The clock then effectively stops, or even runs backwards. In view of these complications we use not a single age but a range, 2–3 MYA, for the arrival of the first finches. It reflects uncertainty in the estimate, and awareness that it is likely to be revised with additional research.

COLONIZATION

Divergence from their mainland relatives began when the finches colonized the Galápagos islands. Where did they come from? Their point of departure from the mainland may have been close to the same latitude. One of the candidate relatives, *T. obscura*, is currently distributed on the South American mainland both north and south of the equator, as the ancestor might have been. A more northerly alternative is suggested by genetic affinities with species currently in the Caribbean (Burns et al. 2002), similar affinities for Galápagos mockingbirds (Arbogast et al. 2006) and other bird species (Swarth 1934), and a few affinities between plants in Galápagos and Mexico, Central America, and the Caribbean (Wiggins 1966, Porter 1976). On the other hand, the climate was warmer and possibly wetter 2–3 MYA, and mainland distributions of tropical flora and fauna may have extended further south to the same latitude as the islands when the mockingbirds and some of the plants are thought to have arrived.

Reaching the archipelago is a remarkable achievement. Situated 900 km from continental Ecuador, the archipelago is a remote place for birds to visit. Colonization is a highly improbable event, and improbable events arise in improbable and hence rare circumstances. What might those circumstances have been?

Any answer must be speculative, even if rooted in current phenomena. Circumstances that promote dispersal of finches *within* the Galápagos archipelago today are high finch density following prolific breeding in El Niño years, and forest fires caused by volcanic eruptions (Plate 5). Therefore perhaps unusual dispersal activity from the mainland was induced by unusual volcanic activity in the Andes and the burning of forests. One can easily imagine large numbers of finches and other birds in coastal regions flying out to sea to escape the heat, flames, and smoke. According to one calculation ancestral Darwin's finches arrived in a moderately large flock (or a few small ones). Modern finches are genetically diverse at the major histocompatibility (Mhc) locus, and Vincek et al. (1996) used the diversity of alleles of a set of genes (class II) to calculate that the original colonists numbered at least 30 individuals. This is one reason for thinking that dispersal was impelled by a major disturbance on the mainland, and was not just the aberrant behavior of a few lost avian wanderers.

THE ECOLOGICAL THEATER

When the ancestral finches arrived they would have encountered a much more limited environment than exists today. Even though the archipelago has been in existence for more than 10 MY, according to evidence from islands now

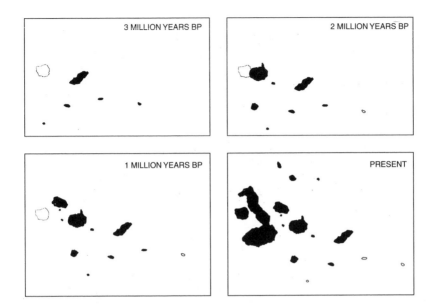

Fig. 2.2 Increase in number of islands since 3 million years ago (3 MYA). Islands that are now completely submerged (seamounts) are shown unfilled. Islands have been formed from a 'hotspot' beneath modern Fernandina, shown by a broken line, or in its vicinity. Reconstructed from data in Christie et al. (1992). From Grant and Grant (1996a).

submerged (Christie et al. 1992, White et al. 1993, Sinton et al. 1996), most of today's islands arose through volcanic activity after the finch ancestors arrived. Three million years ago there were perhaps as few as five islands (Fig. 2.2), and three of those are now submerged (Grant and Grant 1996a). The number of islands subsequently increased as a result of volcanic activity centered on and near a hotspot beneath the western island of Fernandina. The Pacific plate on which the islands sit was periodically punctured as it carried them ESE towards the South American mainland at half a centimeter a year (Sinton et al. 1996). As it did so islands gradually subsided. Old islands were once higher than they are now, and the oldest ones have sunk out of sight.

When the finches arrived ~3 MYA global temperatures (Fig. 2.3) were estimated to be higher than now (see next paragraph), and El Niño conditions are thought to have been permanent (Cane and Molnar 2001, Fedorov et al. 2006). This leads us to believe that the ancestral finches encountered a climate and vegetation on Galápagos more like those of modern-day Cocos

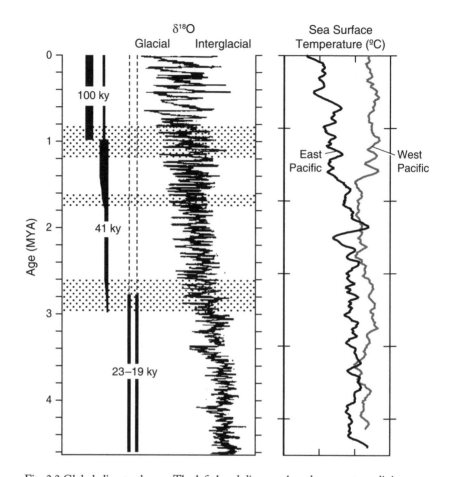

Fig. 2.3 Global climate change. The left-hand diagram, based on oxygen radioisotopes in marine sediments from deMenocal (1995), shows there has been a cooling trend with increasing oscillations between warm and cold periods. Vertical bars on the left-hand side show that the periodicity of the oscillations in thousands of years (ky) has changed abruptly twice, and at ~1.7 MYA their amplitude changed during the 41 ky phase. The right-hand diagram shows a divergence in sea surface temperatures between east and west Pacific starting about 1.7 MYA (from Wara et al. 2005).

island: warmer, wetter, and more humid, fostering rainforest from coast to island peaks (Plate 6). Cocos Island has high temperatures similar to those of coastal habitats in the Galápagos, combined with annual rainfall equivalent to what the Galápagos now receives maximally in an El Niño year at high elevations!

A CHANGE OF SCENERY

Since the arrival of the first finches the Galápagos archipelago has changed in climate more or less continuously and in step with solar rhythms (Zachos et al. 2001). The evidence for environmental change comes more from global indices than from local measurements. Paleoclimates are inferred from deposits of dust particles, plant remains, and the skeletons of plankton in sediments in the sea and lakes, and are dated with radioisotopes (deMenocal 1995). The global climate has cooled in the last 3 MY (Fig. 2.3). A major event 2.73 MYA or slightly earlier was the onset of glaciation in the Northern Hemisphere (Haug et al. 1999, Cane and Molnar 2001), with effects on the climate as far away as the tropics and Antarctica. At about 2.5 MYA or slightly later the Panama isthmus closed for the last time after a brief period of reopening >3.0 MYA, thereby isolating Atlantic and Pacific waters (Cronin and Dowsett 1996). A change in the strength and direction of winds associated with the closure and altered ocean circulation may have been a factor in the colonization of Galápagos, especially if the Panamanian region was the point of departure.

Since then the climate has changed, but not always smoothly. Two changes were especially abrupt. The first occurred approximately 1.7 MYA. It has been attributed to an intensification of temperature fluctuations with a periodicity of 41,000 years under the influence of the tilt of the earth (obliquity) as it rotates around the sun. It resulted in a strengthening in the east-west zonal atmospheric (Walker) circulation and a shift in its position (Trauth et al. 2005). As determined from magnesium/calcium ratios of foraminifera, a difference in sea surface temperatures across the Pacific arose at ~1.7 MYA (Fig. 2.3), or perhaps earlier (Lawrence et al. 2006). Before then conditions were uniformly warm and humid. The difference arose through a decline in temperatures in the east (Wara et al. 2005), and it presaged the modern El Niño–Southern Oscillation (ENSO) phenomenon, with its swings from hot and wet El Niño conditions in the eastern Pacific to cool and dry La Niña conditions, currently at two- to eleven-year intervals.

The second change took place approximately one million years ago when the 41,000-year periodicity switched abruptly to a 100,000-year periodicity under the influence of the eccentricity of the solar path. It resulted in large temperature fluctuations in association with glacial/interglacial cycles, and seasonal aridity. Glacial cooling was accompanied by a reduction in atmospheric carbon dioxide. This favors plants that use the C4 method of photosynthesis (Colinvaux 1996), such as many grasses that exist in open habitats, over those with C3 photosynthesis such as most trees.

THE EVOLUTIONARY PLAY

The initial evolutionary pathway taken by *Certhidea* warbler finches was towards exploitation of small arthropods, soft tissues of fruits, and nectar and pollen from small flowers, in a rainforest-like environment. Finches diversified into several species with more robust beaks (Plate 1) as the number of islands increased (Fig. 2.4), the climate cooled, and the vegetation changed. A literal reading of the phylogeny in a 2–3 MY framework places the origin of the sharp-beaked ground finches and vegetarian finch during the early phase of the gradual intensification of the Pacific temperature gradient and ENSO. The origin of tree finches and most of the ground finches occurred after the archipelago started to become more seasonally arid.

Changes in climate and vegetation were more than just background, they influenced the finch radiation directly. For example, the seasonally arid lowland environments (Plate 7) currently occupied by granivorous (seed-eating) and cactus-exploiting ground finch species apparently did not exist in the early stages of the radiation. Those finch species could only have evolved when the appropriate environment appeared. A changing plant and animal community, the paleocommunity, then guided their subsequent evolution.

RECENT HISTORY

Lying between the short-term oscillations in climate associated with ENSO and the long-term oscillations associated with glacial/interglacial cycles are fluctuations of different scale lengths, including Pacific decadal and multi-decadal oscillations (Miller et al. 1994, Zhang et al. 1998, Chavez et al. 2003), 1,000 years (Raymo 1998, Baker et al. 2001, Turney et al. 2005, Lea et al. 2006), and 20,000 years (Baker et al. 2001). These have affected Galápagos organisms through changes in temperature, precipitation, and sea level.

Changes in sea level have influenced the distance between islands, their connectedness, and even the existence of islands. Sea level was at a minimum

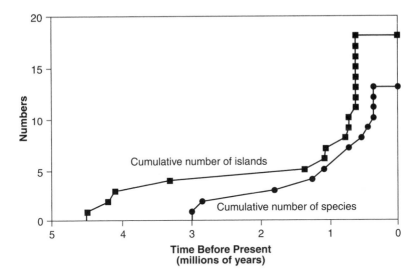

Fig. 2.4 Parallel increase in number of islands and number of finch species. The accumulation of species reflects the results of speciation minus (unknown) extinction; only present-day species could be used to draw the curve. From Grant and Grant (1996a).

about 30,000 YA, ~125m lower than today's, and has risen more or less steadily over the last 22,000 years as revealed from coral cores in Barbados (Bard et al. 1990). The last time sea level was as high as it is now was 120,000 YA (Lambeck and Chappell 2001).

The last Ice Age cooling resulted in a drop of 3–4°C in the equatorial Pacific to a minimum at ~21,000 years ago (Kerr 2001) or somewhat earlier (Lea et al. 2006), and even lower than this (2–6°C) in the previous glaciation ~130,000 years ago (Tudhope et al. 2001). Temperatures were never much higher than today's. Direct evidence of Galápagos climate comes from two sources: analysis of a 135,000 year record of magnesium/calcium and stable oxygen isotope ratios in marine deposits of foraminifera (Lea et al. 2006), and analysis of deposits and pollen in cores taken from El Junco lake on San Cristóbal island (Colinvaux 1972, 1984) and Arcturus lake on the island of Genovesa (Goodman 1972). Lake deposits indicate oscillations in the climate, with periods of dryness alternating with warmer and wetter periods. For example, El Junco lake dried out completely from > 30,000 years ago until about 10,000 years ago in the Holocene (Colinvaux 1984) when temperatures reached a maximum (Lea et al. 2006).

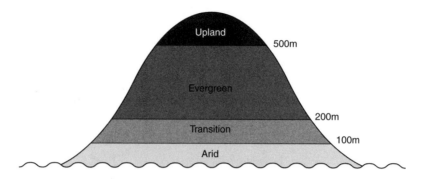

Fig. 2.5 Altitudinal zonation of habitats. Arid and Transition zones are dominated by a variety of deciduous trees and shrubs. In the moist evergreen forest *Zanthoxylum fagara* and *Scalesia* trees are locally dominant. Grasses, ferns, and sedges constitute the Upland zone. Boundaries between zones are not sharp, and their altitudes vary among islands. After Bowman (1963) and Hamann (1981).

Temperatures and precipitation changed from one state to another long enough (> 1,000 years) for vegetation zones to shift (Woodward 1987) and plant evolution to occur in response (Davis et al. 2005). As temperatures rose and fell, temperature- and moisture-dependent altitudinal zonation of plants would have shifted upwards and downwards at different times. The highest and lowest habitats (Fig. 2.5) would have been most vulnerable to constriction and even elimination. On the Mediterranean island of Corsica these habitats have the least number of endemic bird species (Prodon et al. 2002), and similarly on Galápagos these are the habitats occupied by the most recently evolved species: ground finches and tree finches. A change in the moist evergreen forests at high elevations (Plate 7) is indicated by the relative scarcity of endemic plants (Johnson and Raven 1973), although Hamann (1981) found fewer in the Transition zone at lower elevations (Fig. 2.5). Change in the lowest zone is suggested by the anomalous occurrence of plants on some islands with apparently unsuitable climates and soil. For example, the flame tree *Erythrina velutina* occurs at mid-elevations on the high islands of Santa Cruz, Santiago, and Isabela. A single tree of this species occurs on the low and arid island of Genovesa, and less than a dozen occur on the low, arid, and even more remote island of Wolf. They may be the remnants of much larger populations on these islands under cooler, wetter, or less seasonally arid conditions than prevail today.

Thus the last 100,000 years have been a time of change, and although unrecorded, new immigrations and extinctions are likely to have occurred. Finch populations and their habitats did not remain constant in size or distribution, but expanded and contracted. In the Galápagos theatre the scenery has constantly changed, and so have the actors.

SUMMARY

A full understanding of the present requires knowledge of the past. This chapter summarizes the history component of the natural history of Darwin's finches. According to estimates based on a molecular clock applied to mitochondrial DNA data the history began with the arrival of the ancestral species in Galápagos 2–3 million years ago (MYA). The arrival may have been associated with climatic and geophysical changes, because at approximately this time glaciation in the Northern Hemisphere began, the Panama isthmus closed for the last time, and the northern Andes were being uplifted. The first speciation resulted in two groups of morphologically similar warbler finches. One of the groups appears to have given rise to all other species. Early derivatives were the Cocos finch, highland populations of the sharp-beaked ground finch, and the vegetarian finch. Later ones were five species of tree finches and five additional species of ground finches. Not all of the contemporary ecological niches were available when the ancestors arrived. The Galápagos islands have undergone substantial change since then. The number of islands has increased; their sizes, elevations, and distances from each other have changed as a result of fluctuating sea level; climate has changed from warm, wet, and permanent El Niño–like conditions to generally cooler and drier conditions, with strong seasonal and annual fluctuations. Therefore the drama of the finch radiation did not unfold in a single act with unchanging scenery on the Galápagos stage, but in a dynamic environment: numbers and types of opportunities for finch evolution increased as the number of islands increased and the food of finches—plants and the arthropods that feed on them, and on each other—increased in diversity and changed in distribution.

Modes of Speciation

I do not believe that one species will give birth to two or more new
species as long as they are mingled together within the same
district. . . . It would have been a strange fact if I had overlooked the
importance of isolation [in species formation], seeing that it was such
cases as that of the Galapagos Archipelago, which chiefly led me to
study the origin of species.
(Letter to M. Wagner, 1876; Darwin 1887, p. 159)

Among all scientifically tractable questions about speciation, the most
hotly contested concerns its biogeography.
(Coyne and Orr 2004, p. 83)

THE FORMATION OF NEW SPECIES

THE CENTRAL PROBLEM of adaptive radiation, indeed for all discussions
on the origin of biological diversity, is speciation, that is, the question
of how and why one species gives rise to two. For most biologists
"speciation is the process of differentiation within populations and of the rise
of genetic isolation between populations formerly part of the same species"
(Simpson 1953, p. 380). The evolution of many species from one is the result of
a repetition of this basic process. Ecological circumstances undoubtedly differ
from case to case judging from the different products, but the principal ingredients are the same. Those ingredients are a separation of members of a single
population into at least two groups, and a barrier to interbreeding between
those groups. The question of how these ingredients could arise is the subject of
this chapter.

Two Groups from One

In an archipelago setting like the Galápagos, two groups can form from one by individuals of a population flying to an unoccupied island and establishing a new population. In geographical isolation the two populations inevitably diverge as a result of different mutations arising and becoming fixed by chance (Muller 1940). No two environments are the same, therefore the populations will also diverge through natural selection. At some time later when, through further dispersal, members of the two populations come into contact they either attempt to breed with each other and fail, or do not even make the attempt. This is possibly the simplest, as well as the oldest, way of explaining how two coexisting species evolve from one (Grant 2001). Its history was traced back by Mayr (1942, p. 156) beyond Darwin to the early-nineteenth-century geologist Leopold von Buch (1825), who wrote about the Canary island flora as follows:

> The individuals of a species spread out over the continents, move to far-distant places, form varieties (on account of the localities, of the food, and the soil), which owing to their segregation [i.e., geographical isolation] cannot interbreed with other varieties and thus be returned to the original main type. Finally these varieties become constant and turn into separate species. Later they may reach again the range of other varieties which have changed in a like manner, and the two will now no longer cross and thus they behave as "two very different species."

This scheme has come to be known as the allopatric model of speciation. The main elements are (a) a phase of geographical isolation (allopatry) with (b) local adaptation, and (c) a secondary phase of geographical contact (sympatry) with (d) no interbreeding. The outstanding omission is any explanation of how "varieties become constant and turn into separate species." Explanation came later, only after Wallace (1855, 1871) and Darwin (1859) had developed their ideas of evolution by natural selection, Mendel's Laws of Inheritance (genetics) were rediscovered at the end of the century, and Dobzhansky (1937) put them in the relevant framework of population genetics.

The theory of allopatric speciation was adopted by almost all biologists up to the time of the New Synthesis (e.g., Perkins 1913, Grinnell 1924, Rensch 1933, Stresemann 1936, Huxley 1938, 1940, 1942, Mayr 1942), and specifically

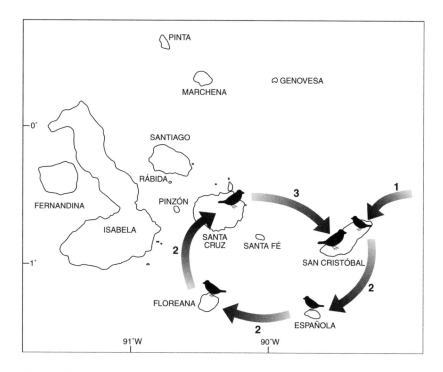

Fig. 3.1 Allopatric speciation, in three stages: initial colonization (1), establishment of a second and additional populations (2), and secondary contact between two populations (3). Choice of islands is arbitrary. Repetition of stages 2 and 3 in other parts of the archipelago gives rise to more species. After Grant and Grant (2002a).

applied to Darwin's finches by Stresemann (1936) and Lack (1947). Figure 3.1 represents the theory as a model, which is an abstraction designed to capture the essence of speciation from a mass of particulars. It illustrates the three steps in Darwin's finch speciation, involving colonization, divergence, and formation of a barrier to interbreeding at secondary contact (Grant 1981a, Grant and Grant 2002a).

DIVERGENCE IN ALLOPATRY

In step 1 an island is colonized by the ancestral species (Fig. 3.1). San Cristóbal, the island arbitrarily chosen to illustrate the initial colonization, is the closest to the continent and also the oldest. The newly established population evolves, by becoming better adapted to the prevailing conditions through natural selection, as well as through some random genetic change. In step 2 a second island is colonized by a few dispersers at some later time, and because the island differs ecologically from San Cristóbal further adaptive change takes place. Geographically separated populations have begun to diverge. The process of allopatric divergence may be repeated several times on several islands before two populations encounter each other in sympatry, arbitrarily chosen in the figure to occur where the cycle of events began, on San Cristóbal (step 3).

COEXISTENCE IN SYMPATRY

Speciation is completed when members of two diverged populations coexist in sympatry without interbreeding. Coexistence is made possible by ecological differences, but the key process by which they "turn into separate species" is the acquisition of reproductive differences that ultimately prevent them from exchanging genes through interbreeding. According to the simplest (and earliest) version of the allopatric model of speciation both ecological and reproductive differences evolve completely in geographical isolation.

Coexistence without interbreeding is not the only possible outcome of step 3. At the opposite extreme members of the two populations freely interbreed, the offspring of mixed matings are fully fit inspite of the genetic differences between their parents acquired in allopatry, and the two populations merge into one. Speciation has collapsed. Absorption of newcomers into the population of residents may be the usual outcome, happening repeatedly, early in speciation.

Between complete separation and complete mixing lies a third possibility: rare or occasional interbreeding with reduced fitness of the offspring thereby produced. The third possibility is interesting because it represents a stage in the process of speciation on the way to complete reproductive isolation of one species from the other. It is also unstable. One outcome is divergence of the two populations driven by natural selection, sexual selection, or both as a result of the lower fitness of hybrids. In this scheme hybridizing individuals are at a selective disadvantage, whereas those with specific traits unique to their group mate with members of their own group without loss of fitness. Those traits and the discriminating use of them constitute barriers to interbreeding (mating). They are thus pre-mating barriers that evolved in allopatry and become strengthened or reinforced by selection in sympatry (Dobzhansky 1937). Reinforcement takes place only if the products of mixed mating, the offspring, are relatively unfit—in other words, if there is a post-mating barrier to interbreeding. This may happen intrinsically as a result of their particular genetic (mixed) constitution, which may reduce their viability or fertility, as Dobzhansky (1937) reasoned, or extrinsically for ecological reasons—they may be poorly adapted to the available feeding conditions.

For species to be formed by a combination of changes occurring in allopatry and sympatry the populations must differ to some minimal extent in mate choice and feeding ecology or habitat use at the time of contact, otherwise speciation will collapse. If the difference threshold is not reached, a gradually increasing rate of mixing of the two populations through interbreeding will cause them to fuse into one, or else one population, most likely the incumbent, will competitively eliminate the other.

All these processes may take a long time to reach completion, and oscillate back and forth on the way.

SYMPATRIC SPECIATION

In the allopatric model two groups form first and then a barrier to interbreeding arises. One alternative to this is a simultaneous evolution of both the barrier and the groups in a single location. A population splits *in situ* into two non-interbreeding groups, and so their evolution occurs entirely sympatrically. A major problem is to understand how the process can get started. How does a population split into two? How do the groups reach, and then exceed, the difference threshold for the formation of two species referred to above? Several mathematical models have been developed to explore the conditions under which this could happen.

In simple language, the basic requirement is an unoccupied niche in the environment that some members of the original population can, and do, exploit. Now that two niches are occupied, their occupants are subject to different regimes of selection, and diverge (Maynard Smith 1966, Gavrilets 2004). Interbreeding of members of the two groups diminishes, even ceases, if individuals choose mates among those with the same niche characteristics as themselves (Kawecki 1997, Higashi et al. 1999, Kondrashov and Kondrashov 1999). Alternatively a population could theoretically split into two when similar individuals compete for food, if the degree of competition between phenotypes is positively related to their similarity, and mating frequency is likewise a positive function of phenotypic similarity (Doebeli 1996, Doebeli and Dieckmann 2000, Gavrilets 2004, Bürger and Schneider 2006). The conditions are restrictive and complex, however, (Bürger et al. 2006). A small amount of interbreeding of the two groups would generally prevent divergence. Two analyses conclude that evolutionary splitting occurs only if frequency dependence and assortative mating are both strong (van Doorn et al. 2004, Bürger and Schneider 2006).

An empirical example illustrates the potential. We raised the question of whether sympatric speciation was in progress on Genovesa (Fig. 1.1) upon finding that two groups of *Geospiza conirostris* males sang different songs, differed in average beak sizes and fed on average in different ways in the dry season (Grant and Grant 1979). The crucial test is whether daughters produced by males of one song group mated with males of the same song group (assortative mating). In the next generation mating was not assortative but random (Grant and Grant 1989). They failed the speciation test. We do not know how the differences in morphology and feeding ecology between the two song groups arose in the first place, but we do know that as a consequence of random mating the differences disappeared, and did not reappear in the following decade.

The models are more easily applied to arthropods with discrete niches, such as host-shifting insects (Feder 1998), than to vertebrates, although some success has been claimed for fish (Schliewen et al. 1994, Barluenga et al. 2006). Among birds, brood-parasitic viduine finches in Africa are the best candidates because nests of the different host species where they lay their eggs are discrete and are the focal point of courtship (Sorenson et al. 2003).

If sympatric speciation has occurred in Darwin's finches it most likely happened (a) on large islands because they are the most heterogeneous, and (b) early in the radiation when niches (resources) were available and few species (consumers) exploited them (Grant and Grant 1989).

PARAPATRIC SPECIATION

Two groups are held together by interbreeding and exchanging genes, and they are pulled apart by natural selection through different ecological factors.

The outcome is dependent upon the balance of the two processes. Allopatric populations may occasionally exchange genes, when birds that disperse from one island to another stay to breed, but surely they do so more rarely than groups that originate sympatrically, and for this reason natural selection is likely to be more effective in causing divergence in allopatry than in sympatry. Intermediate conditions between allopatry and sympatry are experienced by geographically adjacent (parapatric) populations. Differences in the selection regimes to which they are exposed might be strong enough to more than counterbalance the equalizing effects of gene flow between them (Slatkin 1975, Rice and Hostert 1993). If so, the populations could in principle diverge to a point at which they no longer interbreed despite geographically restricted opportunities to do so where the two populations and environments come into contact (Endler 1977, Gavrilets et al. 1998). This model of speciation is halfway between evolution in isolation (allopatry) and in its absence (sympatry).

Nesospiza buntings on the Atlantic island of Tristan da Cunha meet some of the conditions of parapatric speciation theory. Morphological differences between neighboring populations along a steep altitudinal gradient persist in the face of a low level of gene exchange (Ryan et al. 1994).

The plausibility of parapatric speciation in Darwin's finches depends on the strength of different selection regimes along an environmental gradient and the degree to which finches are restricted to one regime of selection. Neither of these is known. The most favorable circumstances for parapatric speciation are altitudinal gradients on the high islands of Isabela, Santiago, and Santa Cruz. These islands are all higher than 800 m, and support altitudinal gradients in vegetation that are recognized as habitats: drought deciduous forest at the lowest elevations, transition zone habitat above this, followed by *Zanthoxylum* and *Scalesia* humid forest and grassland-sedge habitat at the highest elevations (Fig. 2.5 and Plate 7). Morphological differences between populations of the same finch species along altitudinal gradients are minor, however. There is a slight tendency for ground finches at high altitude to be larger than those at low altitude (Grant 1999). Populations of the small ground finch, *G. fuliginosa*, 18 km apart on an elevational gradient on Santa Cruz island were found to differ in bill length in each of four years (Kleindorfer et al. 2006). This is encouraging; however, a difference in abrasion of bill tips may have contributed to the difference: lowland birds feed on harder seeds from the ground than do highland birds.

Testing the Models

The most comprehensive surveys of speciation in organisms, with an emphasis on genetical (Coyne and Orr 2004) and ecological processes (Schluter 2000), have concluded, like Mayr (1963), Futuyma (1998) and many others before, that some form of allopatric speciation is probably the most common mode of speciation in nature. Moreover, for an archipelago setting like the Galápagos, Caribbean, Hawaii, or the islands of southeast Asia the allopatric model is the simplest because the two components of speciation, group formation and reproductive isolation, can be accounted for separately, and group formation is unambiguous. Although speciation could proceed sympatrically or parapatrically in some cases, the necessary conditions for that to occur are more specific. Without completely ignoring the alternatives (e.g., see ch. 9) we will henceforth focus on allopatric speciation as being the most plausible model for Darwin's finches.

There are two ways in which models can be tested: by direct study of their measurable components or by indirect study of their implications. For evolution that has occurred in the past the scope for direct study is limited, and more often we must use inferential information about the past extracted from observations and measurements of contemporary phenomena.

In the following five chapters we explore the main ideas about speciation with data from Darwin's finches. We attempt to answer the question of how a finch species evolves into two populations that are so different from each other in ecological and reproductive characteristics that they are able to coexist with little or no interbreeding. We stand a negligible chance of observing the whole process of speciation under natural circumstances. Nevertheless it is possible to make relevant observations of all the steps in the process. We begin by considering the ecological circumstances of divergence in allopatry and coexistence in sympatry. We then discuss the nature of the barrier to interbreeding, how it evolves, whether there is a fitness cost to interbreeding, and if so what the consequences are.

Summary

According to the standard allopatric model, speciation begins with the establishment of a new population, continues with the divergence of that population from its parent population, and is completed when members of two diverged populations coexist without interbreeding in sympatry. Either the barriers to interbreeding evolve entirely in allopatry, or else their incipient evolution becomes strengthened by divergent selection in sympatry. The same dichotomy applies to ecological differences that facilitate their coexistence: either they evolve entirely in allopatry or they become enhanced in sympatry. Speciation is not inevitable. The process may collapse if two populations interbreed freely with little or no fitness loss of the offspring, or if one competitively excludes the other. There are two alternatives to allopatric speciation. Speciation may occur entirely sympatrically, or along an environmental gradient where two parapatric (adjacent) populations are subject to strongly divergent selection regimes. Conditions for these two alternatives are more restrictive than they are for allopatric speciation.

CHAPTER FOUR

Colonization of an Island

The reduced variability of small populations is not always due to
accidental gene loss, but sometimes to the fact that the entire
population was started by a single pair or by a single fertilized female.
(Mayr 1942, p. 237)

Speciation: The Initial Split

Accoording to the standard allopatric model, speciation begins with the establishment of a new population. If the founding population is large, a long time might elapse before substantial divergence occurs through natural selection and random genetic drift, ultimately yielding a new species. This was the prevailing view of how new species form (Stresemann 1936, Dobzhansky 1937, Mayr 1942) when Lack (1947) conducted his pioneering study of Darwin's finches. Alternatively, if the new population is established by a few individuals, possibly unrepresentative of the parent population (Huxley 1938), the founding event is potentially of great importance for the future evolution of the population because crucial events such as random loss of rare alleles could take place while the population is still small. In this chapter we describe the initial colonization of an island and the genetic and phenotypic fates of individuals over the next 23 years.

Establishment of a New Population

Rapid evolution might occur shortly after a new population is founded as a result of random changes and selection. Genetic variation is reduced due to the small number of colonists and to genetic drift (the random loss of alleles),

35

and inbreeding occurs at a high frequency. Genes interact (epistasis), and the new combinations of interacting genes are subject to selection (Fig. 4.1). It has been suggested that the changes could be profound enough to yield a new species, reproductively isolated from the population that gave rise to it. This is the classical founder effects model of speciation of Mayr (1954, 1992), and is the antithesis of the more gradual speciation in the standard model (Grant 2001). It owes much to the work of Sewall Wright on the genetic structure and dynamics of small populations (Wright 1932).

Mayr used the founder effects model to explain why populations of birds on relatively small islands near continental regions are well differentiated, phenotypically (morphologically), while over a much larger area of the continent or on large islands populations are relatively uniform (e.g., Mayr and Diamond 2001). This version of allopatric speciation is sometimes called peripatric speciation to emphasize the peripheral location of speciation. The essential components are small population size and a substantial genetic change soon after the population becomes established. Genetic change plays the key role. Ecological factors are not negligible but their importance is secondary (Mayr 1992).

Although the nature of genetic change during and after a founding event has been investigated theoretically (Carson and Templeton 1984, Gavrilets and Boake 1998, Hedrick 1998), the evidence for profound change leading to reproductive isolation has frequently been questioned (Barton and Charlesworth 1984, Provine 1989, Barton 1996, Gavrilets and Hastings 1996, Coyne and Orr 2004). A resolution of this conflict requires a study of genetic and phenotypic changes occurring during and after a founder event under natural conditions.

FOUNDER EFFECTS: EXPECTATIONS FROM THEORY

There are four genetic expectations from theory, the first three of which can be tested fairly easily:

- Colonization is random with respect to genotype.
- Inbreeding occurs and average fitness declines, but founders have sufficient genetic variation to survive a period of inbreeding depression.
- There is a loss of alleles through genetic drift.
- Alleles come together in new epistatic combinations that are then subject to natural selection.

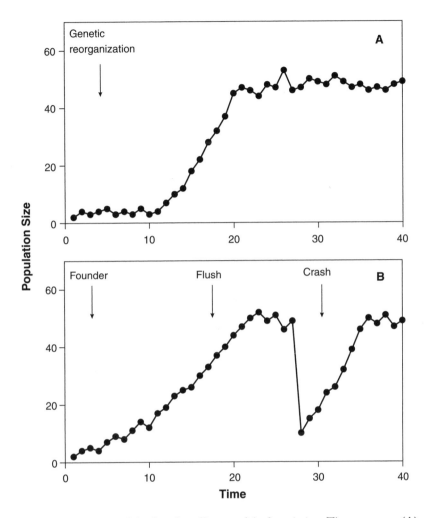

Fig. 4.1 Two versions of the founder effects model of speciation. The upper one (A) shows strong genetic changes taking place soon after a population is founded in a new environment when numbers are low and alleles are lost. Eventually new combinations of surviving alleles are selectively favored (Mayr 1954). In alternative B important genetic changes are initiated by genetic drift as the population increases. New combinations of alleles that would normally be disadvantageous are able to persist under conditions of relaxed (or no) selection. Selection becomes intense at high numbers, or during a subsequent population crash, and could cause rapid morphological change, especially if these events are repeated (Carson 1968). In both schemes genetic change could be strong enough to prevent successful interbreeding with the parent population.

A Colonization Event

Typically a new population is discovered after the initial founding event, and changes can only later be inferred retrospectively. Observing a colonization event requires being in the right place at the right time. We have had that good fortune once, when, towards the end of 1982, a breeding population of the large ground finch (*Geospiza magnirostris*) became established on Daphne Major island. We first describe the context.

In 1973 we began a long-term study of finches on the small island of Daphne Major (Plate 5). The island is centrally located in the Galápagos archipelago, 8 km north of the much larger Santa Cruz island and the same distance west of Baltra (Fig. 1.1). It is a tuff cone with a central crater, 34 ha in area, approximately three-quarters of a kilometer long and 120 meters high. It has never been disturbed by humans. We used mist nets to capture a large number of finches, weighed and measured them, gave each a unique combination of three colored leg bands that were coded to correspond to the number on a fourth, metal, band, and released them for subsequent observation of their feeding behavior (Plates 8 and 9). We increased the samples of identified birds by banding nestlings, then capturing and measuring them when adult. The two principal species on the island are the medium ground finch, *Geospiza fortis* (Fig. 1.3), and the cactus finch, *G. scandens* (Plate 10). Small ground finches, *G. fuliginosa*, are very rare and sometimes absent.

At the end of 1982, just as a major El Niño event was getting underway, the large ground finch, *Geospiza magnirostris* (Plate 10), established a breeding population on the island (Gibbs and Grant 1987a). In preceding years we had observed a few immigrant members of this species on the island in the dry season, but when the rains began they disappeared, presumably returning to their island of origin to breed. Not so in El Niño! Three males and two females stayed to breed. They bred in various combinations and produced a total of 17 fledglings. By 1984 the parents and 14 fledglings had died leaving three offspring, all produced by one female. Therefore the next generation was started by one female breeding with her two brothers. One of the brothers can be ignored because his offspring failed to survive to breed. Thus the population was effectively founded by a single pair, and the next generation comprised a sister-brother pair (Grant and Grant 1995a).

INBREEDING

Some of the expected events occurred soon after the population was founded, most notably inbreeding. The population remained at low numbers for a decade, and only increased with the arrival of another El Niño event of abundant rain in 1991 and in the following two years (Fig. 4.2). As expected from the founder effects model, inbreeding depression, which is considered to be the result of deleterious alleles being brought together in homozygous condition, occurred in the first 10 years (Fig. 4.3). It was more severe than ever recorded in the two resident populations of *G. fortis* and *G. scandens* (Keller et al. 2002). However, survival of the first inbred individuals was unexpectedly high, higher in fact than the survival of all other members of the population in the next 20 years. This complicates the simple expectation of the founder effects model by showing that genetic disadvantages from inbreeding can be overridden by the ecological advantages of abundant food and low density. In addition, the founders were unusually heterozygous at microsatellite loci, hence the first offspring, despite being inbred, were not unusually homozygous.

Also as expected, the population experienced a genetic bottleneck—microsatellite allelic diversity fell (Fig. 4.4). Possibly natural selection on beak morphology occurred at that time, resulting in a small increase in average beak size (Grant et al. 2001).

RECURRENT IMMIGRATION

Most of these observations are to be expected. However, contrary to expectation, the changes we saw in microsatellite allele frequencies were minor. Moreover, one feature was surprising. Immigration did not occur just once but repeatedly, especially in the 1990s. This was surprising because immigration of *G. fortis* and *G. scandens* individuals from other islands and their breeding on Daphne Major appears to be very rare, on the order of one individual per generation at most (Grant et al. 2004), and much too rare to counterbalance the effects of selection (ch. 5). We used statistical tests (assignment tests; Pritchard et al. 2000) applied to microsatellite alleles to identify the island or

39

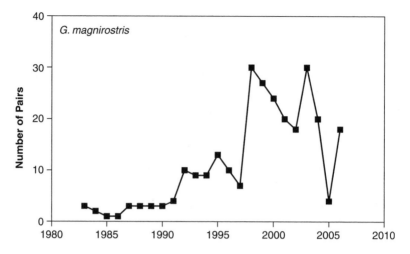

Fig. 4.2 Numbers of pairs of *Geospiza magnirostris* on Daphne Major island. Non-breeding birds, both resident and immigrant, have been omitted.

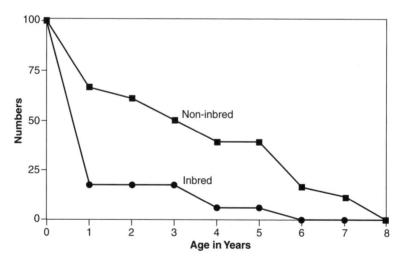

Fig. 4.3 Inbreeding depression in the 1991 cohort of *Geospiza magnirostris* on Daphne Major. Pedigrees and DNA typing of parents and offspring were used to identify inbred birds. Inbred birds survived less well than non-inbred birds in their first year of life. Numbers have been rescaled to 100 for ease of comparison. From Grant et al. (2001).

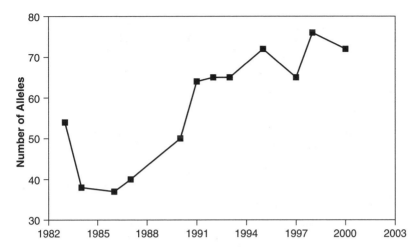

Fig. 4.4 Changes in genetic diversity of *Geospiza magnirostris* on Daphne Major. Numbers of alleles at 16 microsatellite loci have been combined. An initial decline was reversed when additional immigrants contributed new alleles to the population. From Grant et al. (2001).

islands where the *G. magnirostris* immigrants came from. To our (second) surprise we found they came from not one island but four: Marchena, Isabela, Santiago, and Santa Cruz (Fig. 1.1). However, of those individuals that stayed to breed, only one came from the island that is closest to Daphne, Santa Cruz, while all of the rest came from Santiago. Thus the immigrants that stayed to breed were not a random sample of those that arrived; furthermore, they were significantly more heterozygous and larger on average in beak size than the non-breeders.

Whereas most immigrants brought one or two new alleles to the population, one outstanding individual who first bred in 1991 introduced a total of 11 new alleles at the 16 microsatellite loci surveyed. Recurrent immigration rescued the population from its genetic bottleneck, and contributed to an increase in heterozygosity, to the elimination of any long-lasting effects of inbreeding, and to the fact that no substantial change in either phenotypic traits or microsatellite allele frequencies occurred as should have happened according to the founder effects model.

CHAPTER FOUR

AN ALTERNATIVE PHENOLOGY OF FOUNDER EFFECTS

A variation on the founder effects theme was suggested by Carson (1968). He proposed that a population founded by a few individuals initially undergoes some genetic change through random assortment and minor loss of alleles in phase 1. In phase 2 new associations of alleles are formed through recombination and retained as the population increases in size under conditions where natural selection is relaxed or absent altogether. Genetic variation increases. On reaching high numbers the population becomes subject to strong selection, and a crash may ensue. During this third phase of altered combinations of alleles selection may lead to a rapid morphological shift as the population recovers from the crash (Fig. 4.1).

The population of large ground finches followed the prescribed three-phase pattern of founder-flush-crash (Figs. 4.1 and 4.2). During the flush the population increased, with more than 100 young birds being produced in the El Niño year of extensive breeding in 1998. Possibly unidentified immigrants contributed to the increase. From a high point of 350 individuals on the island in 2003, the population crashed to the brink of extinction in 2005, when only four females and nine males were left. The following year the numbers had increased to approximately 45 individuals, partly or wholly as a result of breeding in 2005. However, the population did not change morphologically after the flush at maximum density, nor in the first half of the crash (Fig. 4.5). A few birds with large beaks survived the crash, but so did some with small beaks. The following year, average beak size was no different from what it had been a few years earlier. These are not the expected changes from the founder-flush-crash model of speciation.

Conclusion

These findings are consistent with the general idea that genetic drift and natural selection jointly cause populations to change in the first few generations after a new area is colonized, though not by very much. Some expectations from theory of what happens in the founding of a population from an extremely small number of individuals were confirmed. Alleles were lost and initially the fitness of inbred offspring was low. Unexpectedly, colonization

42

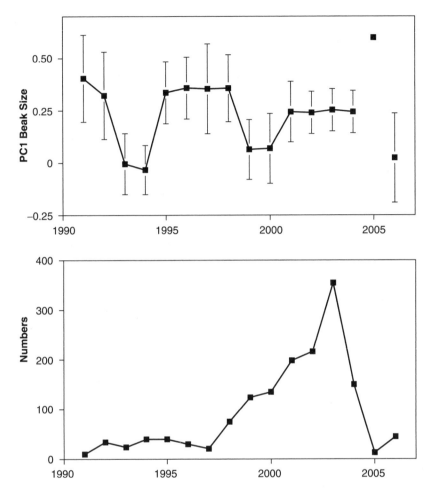

Fig. 4.5 Average beak size in the population of *Geospiza magnirostris* on Daphne Major (upper), during fluctuations in population numbers (lower). Beak size is estimated from a principal components analysis of length, depth, and width measurements. Vertical bars show one standard deviation from the mean. The 2005 sample consisted of only five birds and has a very large standard deviation; all other samples are greater than 10.

was not random with respect to genotype, and immigration occurred repeatedly after the initial colonization and increased the genetic variation.

Overall changes were relatively small in magnitude, for three reasons: selection pressures were weak, the population did not remain small long enough for random genetic drift to be effective, and continuing immigration retarded divergence. If this can be considered representative of how speciation begins, we would have to conclude that it starts slowly, with small steps and not with a single rapid jump. Alternatively, if speciation has occurred through founder events it would have happened on the most isolated islands in the absence of recurrent immigration. Following the establishment of a new population, additional immigration could be the rule rather than the exception.

Therefore we are left with no evidence for founder effects being important in speciation. This does not rule out the possibility that divergence leading to speciation is more likely to proceed in small insular populations than in large ones, as emphasized by Ernst Mayr, nor, as he also emphasized, that most founder events collapse and very few lead to a new species (Mayr 1992). We can envisage one additional way in which founder effects could be evolutionarily significant. Members of a newly founded population might hybridize with another species owing to a scarcity of mates (Grant and Grant 1997a). If this happens, it is possible that a loss of alleles in the bottleneck combined with an addition of a set of new alleles from another species could have a large influence on the composition of the newly founded population and its subsequent evolution (Grant and Grant 1994). Chapter 8 examines the causes and consequences of hybridization.

Species Elsewhere

The founding of a new population is usually missed, and the effects are only sought several generations later (Baker and Jenkins 1987, Baker and Mooerd 1987, Grant 2002). Thus for natural populations, studies are typically retrospective rather than prospective. For example, Fleischer et al. (1991) and Tarr et al. (1998) reported on genetic variation in populations of the Laysan finch 20 years after the birds had been translocated to small islands in the Hawaiian archipelago as a safeguard against the threat of extinction on Laysan itself. Some alleles were apparently lost at microsatellite loci, but there were no major genetic changes. Heterozygosities at microsatellite loci also declined, but unpredictably (Tarr et al. 1998), whereas heterozygosities at allozyme-coding loci actually increased (Fleischer et al. 1991). Clegg et al. (2002a) tested expectations of Mayr's founder effects model with microsatellite loci in a study of silvereyes (*Zosterops*) that had sequentially colonized seven islands near Australia. They found a small, scarcely detectable loss of allelic diversity at each step in the colonization sequence, although hardly on the scale envisaged in the original formulation of the founder effects model. Consistent directional changes towards larger body size at each step in the sequence of colonization events are at variance with expectations from a predominence of random processes in the establishment of new populations (Clegg et al. 2002b).

We conclude that there is no evidence that founder effects contribute significantly to speciation, and that other factors must be of overriding importance.

Summary

Colonization of a new island begins the process of speciation. The process may be very fast as a result of inbreeding, drift, the loss of genetic variation, and selection on substantially altered gene frequencies in new interacting (epistatic) combinations while the population is still small. If the changes are profound enough they could, in theory, yield a new species, reproductively isolated from the population that gave rise to it. A field study of a colonization event on the island of Daphne Major provides little support for this mode of speciation. A new population was founded when two female and three male large ground finches, *Geospiza magnirostris*, established a breeding population at the end of 1982 just as a major El Niño event was beginning. Inbreeding occurred, inbred birds were at a survival disadvantage, and allelic diversity declined. All these were expected from theory. However, overall changes were relatively small in magnitude, for three reasons: selection pressures were weak, the population did not remain small long enough for random genetic drift to be effective, and continuing immigration retarded divergence. If this can be considered representative of how speciation begins, we would have to conclude that it starts slowly, in small steps and without a rapid genetic reorganization.

CHAPTER FIVE

Natural Selection, Adaptation, and Evolution

> When speaking of the formation for instance of a new sp. of Bird
> with long beak Instead of saying, as I have sometimes incautiously
> done a bird suddenly appeared with a beak [particularly] longer than
> that of his fellows, I would now say that of all birds annually born,
> some will have a beak a shade longer, & some a shade shorter, & that
> under conditions or habitats of life favouring longer beak, all the
> individuals, with beaks a little longer would be more apt to survive
> than those with beaks shorter than average.
> *(Darwin 1867; in Burkhardt et al. 2005, p. 299)*

ADAPTATION

IN THE CYCLE of speciation events modeled in Figure 3.1 natural selection is the main driving force in the differentiation of a population in allopatry. Individuals arriving in a new environment encounter new food resources. They vary in traits such as beak size that enable them to exploit the new environment, and some survive and reproduce better than others because of their particular beak sizes. For example, the survivors may have particularly large beaks. This differential process is natural selection, and it occurs within a single generation. If the beak size trait is genetically inherited, the next generation will contain more individuals of the large beak size variant than the previous generation. The change between generations is the evolutionary response to natural selection, and it only occurs if there is genetic variation for the selected trait. Changes are likely to take place not just in the first generation but repeatedly. As a result the population becomes genetically adapted to its new environment.

In this chapter we review the evidence for adaptation to food supplies. The evidence is indirect and direct. The indirect evidence is the repeatedly

46

observed association between beak size and diet among populations of the same species on different islands as well as among populations of different species. The direct evidence is documented evolutionary change in beak size in response to changes in the food environment over time. Continued adaptive evolution requires a continued supply of genetic variation, and we address the question of how that variation is maintained. We conclude the chapter with a discussion of variation in the expression of genes that affect the development of beaks.

Beak Sizes and Diets

Populations of a species on different islands differ in beak size (Fig. 1.3) and shape (Plate 11). Different beaks have been likened to tools of various sizes and shapes that perform different tasks of grasping and crushing (Fig. 5.1). Where beak differences are pronounced we see a correspondence between the average size of beaks and characteristics of the diet such as seed size (Fig. 5.2). The matching variation in both is explicable in terms of (a) seed-cracking performance and (b) variation in food resources available to the finches (Abbott et al. 1977, Smith et al. 1978, Grant and Grant 1980, Boag and Grant 1984a, 1984b, Schluter and Grant 1984a). Thus populations are locally adapted.

Six populations of the sharp-beaked ground finch, *Geospiza difficilis,* supply the clearest example (Schluter and Grant 1984b, Grant et al. 2000, Grant and Grant 2002b). Three of the populations have persisted at mid- and upperelevations in *Zanthoxylum* (cat's claw) forests (Plate 12), on the islands of Santiago, Fernandina, and Pinta (Fig. 5.3); formerly they existed on other large islands, but they became extinct after the forests were destroyed by humans. We believe *Zanthoxylum* forests are old because the sharp-beaked ground finches that inhabit and are largely restricted to them originated early in the phylogeny (Figs. 2.1 and 5.4). The other three populations occupy arid lowland habitat of low islands (Plate 12). If *Zanthoxylum* forest was present on these islands, it disappeared when the climate changed and islands lost elevation as they subsided (ch. 2).

Different populations of this species occupying different habitats feed in different ways on different foods with beaks of different size and shape (Fig. 5.4).

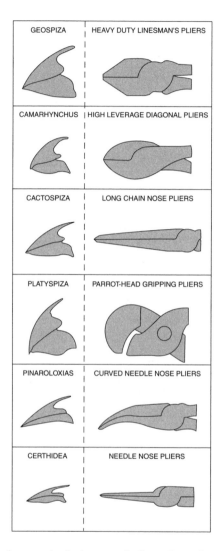

GEOSPIZA	HEAVY DUTY LINESMAN'S PLIERS
CAMARHYNCHUS	HIGH LEVERAGE DIAGONAL PLIERS
CACTOSPIZA	LONG CHAIN NOSE PLIERS
PLATYSPIZA	PARROT-HEAD GRIPPING PLIERS
PINAROLOXIAS	CURVED NEEDLE NOSE PLIERS
CERTHIDEA	NEEDLE NOSE PLIERS

Fig. 5.1 An analogy between beak shapes and pliers. *Cactospiza* species (*pallidus* and *heliobates*) are now considered members of the *Camarhynchus* genus (Table 1.1). From Grant (1999), redrawn by K. T. Grant from Bowman (1963).

On the high islands they have relatively blunt and robust beaks (Plates 2 and 13), and feed on arthropods and mollusks, as well as fruits and seeds in the dry season. On the low island of Genovesa, where they are much smaller in beak and body size (Plate 13), they are more dependent on small seeds, as well as on nectar and pollen from plants including *Opuntia* cactus. The same

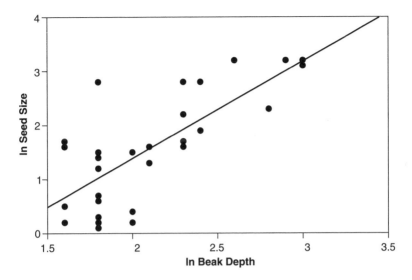

Fig. 5.2 The maximum size and hardness of the seeds that ground finches can crack increases in proportion to their average beak size, shown here on a natural log scale. Based on Schluter and Grant (1984a).

applies to the populations on Wolf and Darwin as well, but with some novel additions.

On the low island of Wolf they exploit seabirds (boobies; *Sula* species) in two dramatic ways (Plate 14). They gain moisture and protein from membranes around the egg as it is being laid. From this simple habit, apparently, there has developed a deeper interest in the egg itself. They kick the egg until it falls or hits a rock and cracks, enabling the finches to open it and consume the contents. Even more bizarre than this, they inflict wounds at the base of wing feathers of the sitting booby and consume the blood. This habit has almost certainly been derived from feeding on hippoboscid flies that suck blood from boobies (Bowman and Billeb 1965), much as mosquitoes suck blood from us. By feeding directly on the boobies' blood the finches have bypassed the evasive flies and shortened the food chain. They have thus added a new element to their feeding niche as a result of a behavioral change. Their long beaks may reflect a subsequent morphological adaptation to this new component, as well as to the task of reaching nectar in *Opuntia* (cactus) flowers. On the neighboring island of Darwin, where their beaks are also long, they are known to feed on eggs (Bowman and Carter 1971) but apparently not on blood, although they do feed on *Opuntia* flowers.

Fig. 5.3 *Geospiza difficilis* populations in the Galápagos archipelago. Open symbols indicate extinct populations. From Grant et al. (2000).

Thus members of the same species of Darwin's finches are versatile in their feeding habits and vary adaptively in beak morphology from island to island. Versatility is fostered by ecological opportunity and driven by food scarcity in the harsh conditions of dry seasons and dry years.

Adaptive Evolution When the Environment Changes

Inferences about evolution in the past derived from examining patterns of variation among islands are underpinned and strengthened by demonstrations of evolution as a contemporary process within a single island. Such demonstration

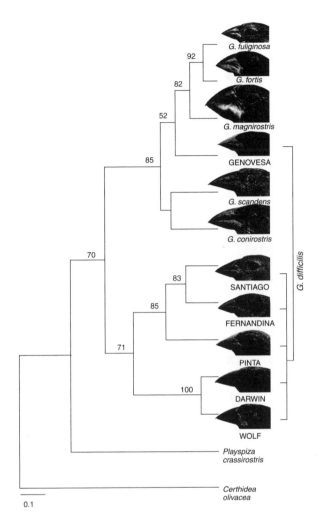

Fig. 5.4 Phylogeny of the sharp-beaked ground finch, *Geospiza difficilis*, estimated from microsatellite DNA comparisons. Names of island populations of *G. difficilis* are in capitals. One branch gave rise to all other species in the ground finch genus. The population on Genovesa is allied with other members of *G. difficilis* in beak shape, plumage, and song (ch. 7). Genetically it is more similar to derived species, perhaps partly as a result of hybridization (ch. 8). See also discussion of this important segment of the phylogeny in ch. 10. A scale of genetic difference is given at the bottom. From Grant et al. (2000).

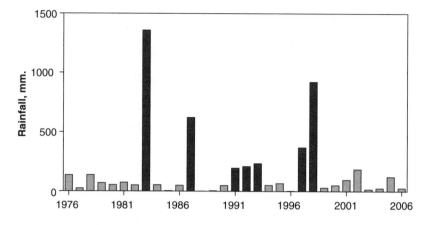

Fig. 5.5 Annual rainfall on Daphne Major. Years of El Niño events are indicated by dark shading. Four other, weaker, events, as classified by sea surface temperature anomalies across the Pacific, occurred during this period (McPhaden et al. 2006). They were not accompanied by extensive rain in Galápagos.

is possible in the Galápagos owing to strongly fluctuating annual rainfall (Fig. 5.5), which affects food supply and finch survival (Boag and Grant 1981, 1984b). The process of evolutionary change has been studied on Genovesa over a period of 11 years (Grant and Grant 1989), and in more detail on Daphne Major for three times as long (Grant and Grant 2002c). By virtue of its small size (0.34 sq. km), moderate degree of isolation (8 km), and approachable finches, Daphne Major (Plate 5) is ideal for studying Darwin's three essential ingredients of adaptive evolution: variation, inheritance, and selection. This has been accomplished by simply capturing, measuring, and banding many finches (ch. 4) to determine phenotypic variation, comparing offspring with their parents to determine inheritance, and following their fates in their isolated world to detect selection.

NATURAL SELECTION

In 1977 we were fortunate to witness a severe drought on Daphne Major. Not so fortunate for the medium ground finches (*G. fortis*), as 85% of them died! At the beginning of the drought these granivorous birds were observed to feed on the small and soft seeds then plentifully available (Plates 15 and

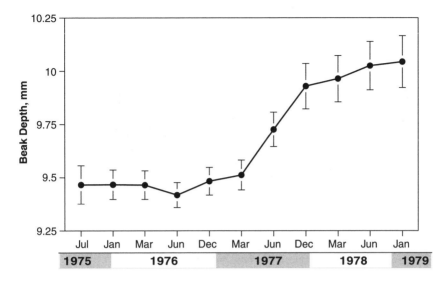

Fig. 5.6 Changes in beak depth in the medium ground finch, *Geospiza fortis*, on Daphne Major as a result of natural selection. Round symbols represent means, vertical bars indicate one standard error. Sample sizes of males and females combined varied from 640 in June 1976 to 61 in January 1979. Adapted from Boag (1981).

16). The drought prevented the regeneration and regrowth of most of the seed-producing plants (Plate 17), and seeds in the soil were eaten and their numbers declined, so finches turned increasingly towards the large and hard seeds, now in comparatively high abundance. Large finches with relatively large beaks and associated powerful muscles can crack the hard woody tissues protecting the seeds of *Tribulus cistoides* (caltrop, Plate 18) with relative ease. Smaller birds, lacking the mechanical power of larger birds (Bowman 1961, Herrel et al 2005a, 2005b), can crack them only with time-consuming difficulty, if at all (Grant 1981b, Price 1987), and they died at a higher rate than large birds. As a result of this differential mortality, average beak size of the population as well as average body size increased, and continued doing so until the rains resumed at the beginning of 1978. Natural selection had occurred (Fig. 5.6), and in a manner remarkably close to Darwin's reasoning at the head of this chapter.

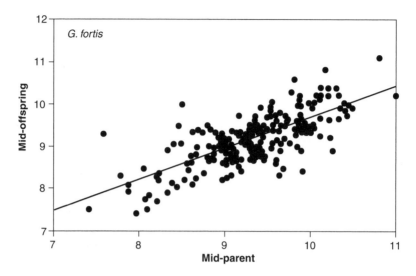

Fig. 5.7 The relationship between *Geospiza fortis* parents and offspring in beak depth, on Daphne Major. Average values in millimeters of offspring in each family (mid-offspring) are shown in relation to the average values of their parents (mid-parent). The slope of the least squares regression line (0.74) is an estimate of the heritability. From Grant and Grant (2000).

EVOLUTION

Evolution—a change from one generation to the next—only happens if the selected trait is heritable; natural selection without heritable variation does not produce an evolutionary change. The strong association between the measurements of beak and body size of fullgrown offspring and their parents (Fig. 5.7) shows that these traits of medium ground finches (*G. fortis*) are indeed highly heritable (Boag 1983, Keller et al. 2001). Similarly high heritabilities have been found in cactus finches (*G. scandens*) on Daphne Major (Grant and Grant 2000) and large cactus finches (*G. conirostris*) on Genovesa (Grant and Grant 1989).

Given the strong inheritance of beak size, directional natural selection in 1977 should have resulted in measurable evolutionary change in the next generation of *G. fortis*. It did. Offspring of the survivors had larger beaks, on average, than did the population before natural selection occurred. This measure of evolution matches closely the value predicted by a formula that animal breeders use: that is, the heritability multiplied by the selection differential, which is difference between the mean value of a character before and after selection (Grant and Grant 1993, 1995b).

Figure 5.8 helps to visualize the components of adaptive evolution. The upper panel demonstrates natural selection. It is the difference in average beak depth between the survivors (in black bars) and the total population before the drought. Inheritance of beak characteristics is demonstrated in Figure 5.7. Evolution takes place from one generation to the next, and this is shown by a comparison of the full-grown offspring hatched in 1978 in the lower panel with the previous generation at the top before the advent of selection. Birds of the next generation, like their parents who had survived the drought, had large beaks.

OSCILLATING DIRECTIONAL SELECTION

Evolution in response to natural selection is not restricted to one trait or one species, or to one island. We documented the occurrence of natural selection repeatedly on Genovesa in the population of large cactus finches (*G. conirostris*) (Grant and Grant 1989). On Daphne Major it occurs frequently and varies in direction and strength according to the particular set of environmental conditions (Price et al. 1984, Gibbs and Grant 1987b, Grant and Grant 2002c).

The most ecologically dramatic change took place in 1983, the year of an exceptional El Niño event that has been described as the most severe in 400 years on the basis of coral core records (Glynn 1990). *Tribulus cistoides* (Plate 18), the plant that was so crucial to the survival of finches with large beaks in the drought of 1977, was smothered by extensive growth of other plants, especially *Merremia aegyptica* vines. The rains and rampant plant growth continued for 8 months, and even in the following years the effects of El Niño could easily be seen (Plate 19). By the smothering of *Tribulus* plants and cactus bushes, and by the prolific growth of 22 species of plants bearing small seeds, El Niño converted the island from a predominantly large-seed environment to a small-seed environment. Under these altered conditions small-beaked birds had a selective advantage, especially those with pointed beaks (Fig. 5.9), when the island entered the next drying-out episode during the drought of 1985. The direction of evolution had been reversed.

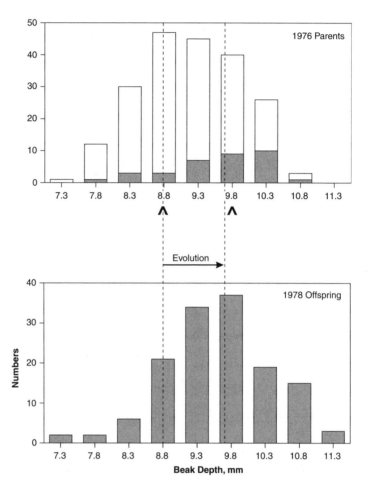

Fig. 5.8 Evolutionary change in beak depth in the population of medium ground finches, *Geospiza fortis*, on Daphne Major. The upper panel shows the frequency of beak depths in the breeding population in 1976, with the survivors of the 1977 drought that bred in 1978 indicated in grey. The lower panel shows the frequency distribution of the next generation produced in 1978. The difference in mean beak depth before and after selection is shown by the pair of carets. An evolutionary response to selection is measured by the difference in mean beak depth between generations: between the population in 1976 before selection and the full-grown offspring produced in 1978, as shown by the arrow. From Grant and Grant (2003).

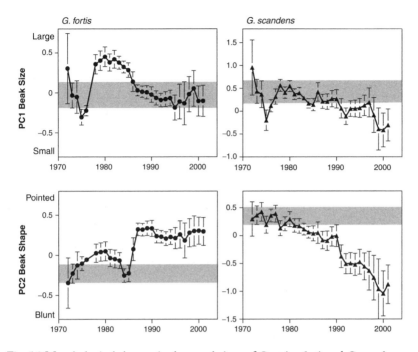

Fig. 5.9 Morphological changes in the populations of *Geospiza fortis* and *G. scandens* on Daphne Major over 30 years. Estimates of mean values are indicated by circles or triangles, and 95% confidence intervals on the estimates are shown by vertical lines above and below the means. The grey zone marks the confidence limits for first large samples in 1973. In the absence of change the mean values should have remained within the zone, but clearly they did not. PC refers to principal components obtained from a statistical analysis of size and shape. From Grant and Grant (2002c).

Oscillating directional selection continued over a period of three decades. Both *G. fortis* and *G. scandens* changed repeatedly in body size, beak size, and beak shape: their environment changed frequently, and so did they. Remarkably, as a result of repeated episodes of natural selection, the finches on Daphne Major are not morphologically the same as their ancestors 30 years ago (Fig. 5.9). *G. fortis* is smaller in average body size and has a more pointed bill than it did in 1973, whereas *G. scandens* is now smaller in average body size and has a blunter beak than it did in 1973.

Extrapolating from Short to Long Term

In the short term of a few decades, oscillations in the direction of natural selection and their evolutionary effects may cancel out, leaving the population at a dynamically equilibrial beak size (Price et al. 1984). On the other hand, over the long term of many decades, centuries, or longer there could be a net trend towards a larger or smaller overall beak size, or more pointed or blunt beak shape, depending on long term climatic trends (ch. 2) and their effects upon the vegetation and food supply.

The main point we wish to emphasize is the evolutionary lability of the studied species, within fairly broad limits. Populations are adaptable. They are capable of undergoing adaptive evolution when their environment changes, either in time, on the same island, or in space, on a new island. Evolution of differences in beak size among populations of the sharp-beaked ground finch (*G. difficilis*) in space is explicable in terms of the adaptive change demonstrated in the medium ground finch (*G. fortis*) in time. A key factor in their adaptability is the large amount of heritable variation they possess.

The Sources of Variation

Natural selection tends to reduce phenotypic variation in a population (Endler 1986). New variation is produced each generation by sexual reproduction when genes come together in new combinations: they are recombined. It is possible that some of the resulting phenotypes will be more extreme than any of those observed before selection, especially if selection is repeated in the same direction and if extreme individual mates assortatively with extreme individual. Such variation beyond the original morphological limits of the population allows the continued directional evolutionary change that must have occurred repeatedly in the adaptive radiation.

Genetic variation is increased by addition from three sources. First, members of another population of the same species differing from it genetically to a small extent may immigrate and breed with local individuals (gene flow). Second, the local birds may breed with members of another species living on the same island or immigrating from another (hybridization). These possibilities will be discussed in chapter 8. Third, the ultimate source of all new genetic variation is mutation, the chemical change of genes.

Insights into how changes in beak morphology occur can be derived from studies of beak development. One can think of natural selection on variation in adult morphology as being indirectly selection on variation in the developmental programs that give rise to adult form. Thus identifying differences between species in the expression of individual genes involved in beak development allows us to infer the effects of recombination and mutational change. Populations of different ground finch species, and to a lesser extent populations of the same species, differ along two axes of variation in their beaks: depth and width is one, length is the other. Genes involved in the pathways along each of these two axes have been identified. They are expressed differently in different species.

How Beaks Are Formed

Beaks develop early in embryonic life. At about the third or fourth day mesenchyme derived from the neural crest gives rise to skeletal projections of upper and lower mandibles (Fig. 5.10). The process of beak formation at this early stage is controlled by two signaling molecules called fibroblast growth factor 8 (Fgf8) and sonic hedgehog (Shh). They are expressed in adjacent areas of the epithelium that covers the mesenchyme. The domain of expression of the *Fgf8* gene is the dorsal fronto-nasal primordium (FNP) and the ventral mandibular nasal primordium (MNP). The intervening region is the domain for *Shh*. The two molecules induce cartilage outgrowth where the domains meet (Abzhanov and Tabin 2004, Wu et al. 2004). This is the origin of the beak.

Depth and Width

The two signaling molecules involved in beak initiation also synergistically induce expression of other factors in the underlying neural crest mesenchyme. These factors include the signaling molecule *Bmp4* (bone morphogenetic protein 4), which is important in the origin of beak shape differences between species.

59

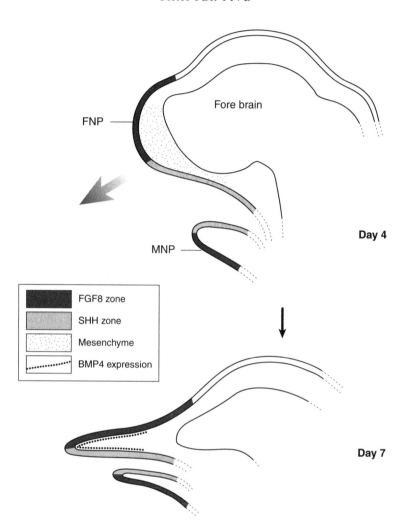

Fore brain

FNP

MNP

Day 4

FGF8 zone
SHH zone
Mesenchyme
BMP4 expression

Day 7

Fig. 5.10 Development of a beak. Two signaling molecules, fibroblast growth factor 8 (Fgf8) and sonic hedgehog (Shh), induce cartilage outgrowth (shown by thick arrow) where the domains meet in adjacent areas of the epithelium. The *Fgf8* domain is the dorsal fronto-nasal primordium (FNP) and the ventral mandibular nasal primordium (MNP). The intervening region is the domain for *Shh*. Expression of another signaling molecule, bone morphogenetic protein 4 (BMP4), is also induced in the mesenchyme of this region.

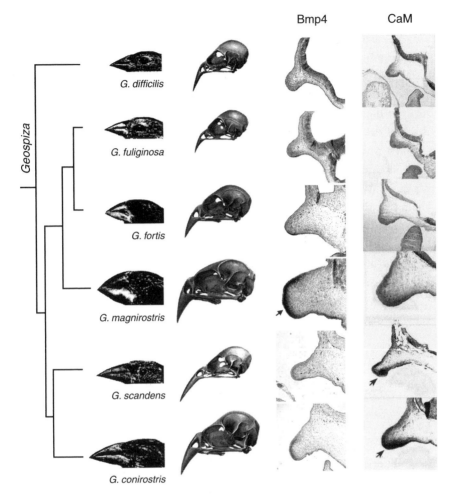

Fig. 5.11 Different expression of *Bmp4* and *CaM* genes in species of ground finches at day 5 of embryonic growth. Notice the strong and early expression of *Bmp4* in the species with the largest beak, *G. magnirostris*. At the same time in development *CaM* is expressed strongly in the species with pointed beaks, *G. conirostris* and *G. scandens*. From Abzhanov et al. (2004, 2006).

At day 5 the expression of the *Bmp4* gene is detectable at low levels in the mesenchyme of *G. difficilis* and other species, but at a dramatically higher level in the largest species, *G. magnirostris* and to a lesser extent *G. conirostris* (Fig. 5.11). At days 6–7 *Bmp4* expression is elevated in all ground finch species but still only weakly in those (*G. difficilis* and *G. scandens*) that have relatively long and shallow beaks (Fig. 5.4). The functional role of *Bmp4* in affecting beak shape development has been experimentally demonstrated in chickens. It is possible to misexpress the gene by injecting it with a retroviral vector in the distal mesenchyme of the upper beak at day 6. This mimics the natural occurrence of elevated levels of *Bmp4* at the same stage in *G. magnirostris*. When this was done it produced very *magnirostris*-like beaks, both in width and in depth of the upper mandible (Abzhanov et al. 2004).

Bmp4 gene expression occurs earlier in *G. magnirostris* than its relatives, more intensely and over a wider area of the developing beak. As a result, its beak becomes deep and broad. We suspect the gene itself does not vary much among species because it is involved in several basic metabolic processes and is expressed at different stages of development in several organs and tissues of vertebrates. Instead, changes in the regulation of the gene, including mutation, may have been responsible for changes in adult beak size and shape in finch populations. There are many more components of the network of interacting factors that govern the development of beaks. One of these is known: it influences the length of beaks.

Length

A second pathway, involving expression of the gene *calmodulin*, influences beak length (Abzhanov et al. 2006). The calmodulin (CaM) molecule is involved in calcium signaling. It is expressed at higher levels at the tips of the long and pointed beaks of the two species of cactus finches than in the more robust beaks of *G. fortis*, *G. magnirostris*, and *G. difficilis* embryos at the same stage of development (Fig. 5.12). Calmodulin has been detected at about the same time in embryonic development as differences in *Bmp4* expression are becoming clear (Fig. 5.11), but is apparently independent of *Bmp4* expression. Therefore changes affecting the regulation of the CaM-dependent pathway molecule, possibly involving spontaneous mutation, are likely to have provided the important selectable variation in adult beak morphology of the two long-beaked cactus finch species.

This finding, in combination with the discovery of the *Bmp4* gene, is an exciting beginning to the complex task of unraveling the genetic changes that occurred in the adaptive evolution of Darwin's finches. It takes many genes to make the beak of a finch, and we would like know how many change when a new species evolves. Research in this area is continuing.

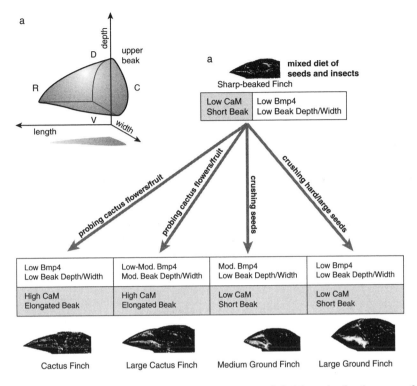

Fig. 5.12 Summary of the effects of two genes, *Bmp4* and *CaM*, on the development of finch beaks. From Abzhanov et al. (2006).

SUMMARY

The main driving force in the allopatric differentiation of populations is natural selection. There are two types of evidence for natural selection: patterns and process. The patterns are associations between beak morphology, food availability, and diets. The clearest example of allopatric differentiation is provided by the six populations of the sharpbeaked ground finch, *Geospiza difficilis*. Different populations of this species occupy different habitats, and feed in different ways on different foods with beaks of different size and shape. They are locally adapted. Natural selection as a process has been demonstrated repeatedly in studies of ground finches on Daphne Major and Genovesa. Selection has been shown to occur frequently when the environment changes, and to vary in direction and strength according to the particular set of environmental (food supply) conditions. Selection on beak size variation arises from differences in mechanical efficiency: the larger the beak and associated musculature a bird has, the larger and harder are the seeds that can be cracked open. It results in evolutionary change because the morphological traits subject to selection are heritable. Natural selection on variation in adult morphology is indirectly selection on variation in the developmental programs that give rise to adult form. Recent molecular genetic analyses of ground finches (*Geospiza* spp.) have identified two signaling molecules that are involved in the development of deep and wide beaks on the one hand and long beaks on the other hand. Their different regulation among species appears to be the key to understanding the genetic foundations of adaptive evolution.

CHAPTER SIX

Ecological Interactions

To me this specialization [of bill form] indicates the severe
competition that has taken place between the Drepanids
in past ages.
(Perkins 1903, p. 302)

INTRODUCTION

A CRUCIAL STAGE in the cycle of speciation events is the establishment
of sympatry by two previously separated populations derived from the
same ancestor (Fig. 3.1, step 3). They make contact through dispersal
of individuals of one population to an island occupied by another. Whether
they interact or not is determined by how different they are at that time. If
they are not very different they may compete for food resources, and if this
happens one population might go extinct, or natural selection might cause
resource-exploiting traits such as beak size or beak shape to diverge, resulting
in diminished competition between them. This latter process, known as character displacement (Brown and Wilson 1956, Grant 1972), facilitates long-term
coexistence in sympatry.

In this chapter we describe what is known about ecological competition
between Darwin's finch species at their initial encounter. In the next two
chapters we consider how mates are chosen (ch. 7), interbreeding (ch. 8), and
whether there is selection against those that interbreed (hybridize). Interspecific
competition is further discussed more generally in relation to adaptive radiation in chapters 10–12. Here the focus is on the role of ecological competition
in speciation.

Competition

How can the signature of divergence as a result of competition be detected? It is not enough to observe ecological differences between coexisting, closely related species because those differences may have evolved entirely in allopatry, a necessary precondition for coexistence rather than a consequence of it. However, if the species competed at the time of secondary contact and then diverged, they should be more different where they occur together, in sympatry, than in allopatry. The first part of the chapter discusses this kind of evidence for past competition inferred from population comparisons. Until recently it was the only evidence available, but in 2005 we were able to directly observe the process under natural conditions, and this is described and discussed in the second half of the chapter.

Patterns of Coexistence

David Lack (1947) used the method of comparing sympatric and allopatric populations to draw attention to several patterns of coexistence that he interpreted as evidence of evolutionary effects of competition.

1. *G. difficilis* (sharp-beaked ground finch) is a highland species in the presence of the small ground finch *G. fuliginosa* (Santiago, Fernandina, Pinta) but a lowland species in its absence (Genovesa, Wolf, Darwin). *G. fuliginosa* is always a lowland species, even if not restricted to lowlands. On the three low islands lacking *G. fuliginosa*, *G. difficilis* have less robust beaks than on the other islands (Plates 12–14). Their beak shapes indicate a feeding niche more characteristic of not only *G. fuliginosa*, especially on Genovesa, but also the absent cactus finch *G. scandens*.

2. *G. fortis* (medium ground finch) and *G. fuliginosa* are more similar to each other in beak and body size in the absence of the other, on Daphne Major (only *G. fortis* present, except for occasional and rare *G. fuliginosa* immigrants; see Fig. 1.3) and Los Hermanos (only *G. fuliginosa* present), than in the presence of each other (many islands).

3. *G. conirostris* (large cactus finch) is intermediate in beak form and body size between *G. magnirostris* (large ground finch) and *G. scandens* in the absence of both (Española), and more like *G. scandens* in the presence of *G. magnirostris* (Genovesa) (Plate 20).

Diets Inferred from Beaks

These three examples show that a minimum ecological difference between similar species is necessary for coexistence. If differences are less than the minimum, one species, the first to occupy an island, is presumed to competitively exclude the other. Consistent with a minimum difference, all populations of coexisting ground finch species differ by at least 15% in at least one beak dimension (Grant 1999). Beyond that minimum the greater the difference in beak sizes of two species, the greater is their dietary difference (Fig. 5.2).

Field studies since Lack's visit to the Galápagos in 1938–39 have provided ample quantitative evidence to support the inference of diets from particular beak morphologies. *G. difficilis* feeds like the granivorous *G. fuliginosa* when it is in the lowland habitat of *G. fuliginosa*, and has a different diet that is richer in arthropods when it is in the highlands (Schluter and Grant 1984b). The diet of *G. fortis* on Daphne Major similarly reflects its small *G. fuliginosa*-like beak (Boag and Grant 1984a, 1984b). The diet of *G. conirostris* is more like that of the granivorous *G. magnirostris* (Plate 21) in its absence on Española than in its presence on Genovesa, where *G. conirostris* feeds largely on *Opuntia* cactus products as does *G. scandens* elsewhere (Grant and Grant 1982) (Plates 22 and 23). Sizes of insects eaten by tree finch species tend to covary with beak size (Bowman 1961), possibly to an extent that is dependent on the presence of competitors. This deserves more study.

Interpreting the Patterns

Finding enhanced differences in sympatry implies divergence there: competition, natural selection, and morphological change. But are the pairs of species phylogenetic sisters? If they are, the differences are directly relevant to the question of what happens at the secondary contact phase of speciation. If they are not, but the species are moderately close relatives nonetheless, they are still valuable for illuminating the interactions when species encounter each other.

In the case of *G. fuliginosa* and *G. difficilis* the phylogeny is not well enough resolved to answer the question of relatedness—in fact it is highly confusing (Fig. 2.1; see also ch. 10)—but the answer appears to be no. In the other two cases, *G. fortis* and *G. fuliginosa* are sister species and *G. conirostris* (Española) originated earlier than *G. magnirostris*.

The next question is this: did the state we now observe in allopatry precede or succeed the observed sympatric state? This question is difficult to answer

67

because allopatric and sympatric populations are too similar genetically to enable us to trace the pathways from one to the other. Again the answers are mixed. *G. fortis* and *G. fuliginosa* are allopatric on just two islands right in the center of the archipelago, in contrast to their sympatry on numerous islands. A parsimonious explanation of their history is that allopatric populations were derived from sympatric ones, not vice versa. In fact Daphne Major was part of Santa Cruz island during the last glacial maximum, when sea level was much lower than at present, and only became an island about 15,000 years ago (Grant and Grant 1998a). With the budding off of the island, a population of *G. fortis* was budded off as well. Presumably at some time a population of *G. fuliginosa* there became extinct.

We interpret the phylogenetically nested relationships of *G. conirostris* (Española) and its allies (Fig. 2.1; microsatellites) as evidence of *G. conirostris* on Española being older than all the others. If this is correct, allopatry of *G. conirostris* on Española preceded sympatry with *G. magnirostris* on Genovesa, in which case evolutionary change on Genovesa is evidence of character displacement from the competitor *G. magnirostris*.

Character Displacement and Release

Character displacement is the process of divergence of resource-exploiting traits in sympatric closely related species, and it results in a reduction of interspecific competition (Grant 1972). The opposite, convergence of resource-exploiting traits in allopatry, is character release; in the absence of a competitor species the resource-exploiting trait of each one changes in the direction of the missing species. When Brown and Wilson (1956) first introduced the term "character displacement" they used the example of *G. fortis* and *G. fuliginosa* as a classical case. In fact it is better described as an example of character release (Boag and Grant 1984b). Character displacement is the relevant process in speciation at step 3 of the allopatric model. It has been observed and documented on Daphne Major (Grant and Grant 2006a).

Character Displacement Observed

A breeding population of *G. magnirostris* was established on Daphne Major by two females and three males in late 1982 (ch. 4). This set up the potential for character displacement in *G. fortis* because they both feed on *Tribulus* seeds (Plate 24), albeit with different efficiencies and frequencies. On average large-beaked members of the *G. fortis* population take three times longer than *G. magnirostris* to crack or tear open the woody protection of the fruits (meri-carps) to gain a seed reward (Grant 1981b). *G. magnirostris* compete with *G. fortis* by physically excluding them from *Tribulus* feeding sites, and by reducing the density of *Tribulus* fruits to the point at which it is not profitable for *G. fortis* to feed on them (Price 1987). By depleting the supply of *Tribulus* fruits *G. magnirostris* are predicted to cause a selective shift in *G. fortis* in the direction of small beak size when food supply in general is severely reduced.

The predicted character displacement occurred 22 years after the breeding population of *G. magnirostris* was founded (Fig. 6.1). Initially the population was too small (Fig. 4.2) in relation to the food supply to have anything but a mild competitive effect upon *G. fortis*. Their numbers gradually increased as a result of local production of recruits augmented by additional immigrants, and reached a maximum of about 350 in 2003 (Fig. 6.2). Hardly any rain fell in 2003 and 2004: 16 mm in 2003 and 25 mm, or one inch, in 2004. There was no breeding in either year, little or no renewal of their food (seed) supply, and numbers of both species declined drastically. At the beginning of 2004 *G. magnirostris* numbers (150) were closer to those of *G. fortis* (235) than they had ever been, and being almost twice the size of *G. fortis*, individually, their population biomass was actually about the same as *G. fortis*. From 2004 to 2005 the numbers of both species fell yet further, and *G. fortis* experienced strong directional selection against those individuals with large beaks (Table 6.1). To illustrate with an example, all 10 males with the largest beaks died. Overall beak size rather than beak length was the most important factor that distinguished survivors from non-survivors in each year; an index of beak size was a selected trait in both sexes, whereas beak shape, that is length in relation to depth and width, was not selected in either.

The Competitive Role of G. magnirostris

How do we know, in the absence of experimental evidence, that competition with *G. magnirostris* caused selective mortality in the population of *G. fortis*? Just as in 1977, the cause of natural selection is inferred from correlated evidence: from the limited amount of available food, overlap in their diets, and the direction of natural selection under contrasting conditions.

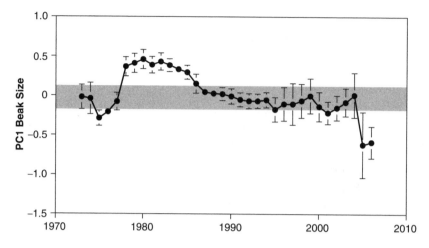

Fig. 6.1 The long-term morphological trajectory of the medium ground finch *G. fortis* on Daphne Major, showing character displacement of beak size in 2004–05 caused by competition with the large ground finch *G. magnirostris*. Mean beak size (solid circles) and 95% confidence intervals (vertical bars) are shown for each year. The grey zone marks the limits of confidence for the first sample in 1973. From Grant and Grant (2006a).

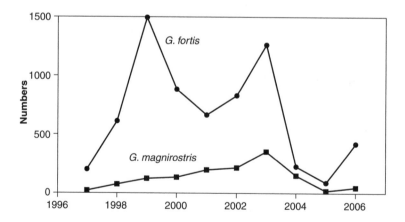

Fig. 6.2 Numbers of *G. magnirostris* and *G. fortis* on Daphne Major before, during, and after a crash. See also Fig. 4.2. From Grant and Grant (2006a).

TABLE 6.1

Selection coefficients associated with two episodes of strong, but opposite, selection experienced by *Geospiza fortis* in the presence (2004) and absence (1977) of *G. magnirostris*.

Statistical significance (*t*-tests) at $P < 0.05$, < 0.01, < 0.005 and < 0.001 is indicated by *, **, *** and **** respectively. PC refers to principal component, a measure of overall body size, beak size, or beak shape. From Grant and Grant (2006a).

	1977		2004	
	Males	*Females*	*Males*	*Females*
Weight	0.88****	0.84***	−0.62*	−0.63
Wing length	0.47***	0.71**	−0.66*	−0.60
Tarsus length	0.24	0.27	−0.48	0.01
Beak length	0.75****	0.88***	−1.08****	−0.95*
Beak depth	0.80****	0.69*	−0.94***	−0.91*
Beak width	0.71****	0.62*	−0.87***	−0.81*
PC 1 body	0.69****	0.73**	−0.67*	−0.52
PC 1 beak	0.80****	0.74**	−1.02****	−0.92*
PC 2 beak	0.23	0.29	−0.34	−0.26
Sample	164	55	47	24
Prop. survivors	0.45	0.42	0.34	0.54

G. magnirostris reduced the supply of *Tribulus* seeds faster than *G. fortis*: an individual *G. magnirostris* consumes as much *Tribulus* food as is needed by two *G. fortis* individuals feeding on nothing else in the same amount of time. As a result of their joint reduction in seed biomass *G. fortis* fed on *Tribulus* in 2004 only half as frequently as in other years (Fig. 6.3). Food supply was not quantified; nevertheless, food scarcity is evident from exceptionally low feeding rates of *G. magnirostris*. In 2004 a minimum of 90 individuals were observed foraging for *Tribulus* mericarps for 200–300 seconds, and none obtained more than one or two mericarps with seeds, whereas under the more typical conditions prevailing in the 1970s a total of eight birds observed for the same length of time fed on 9–22 mericarps, with an average interval between successive mericarps of only 5.5 seconds. The difference in feeding rate is enormous.

Numbers of *G. fortis* declined to a lower level (83) in 2005 than at any time since the study began in 1973. Numbers of *G. magnirostris* decreased by the

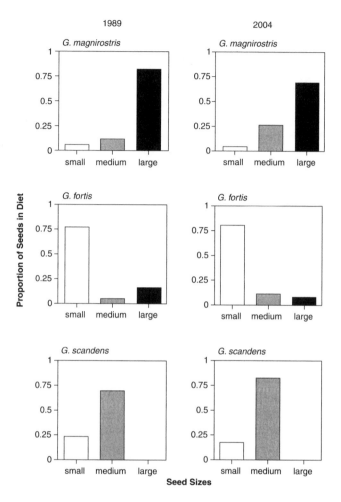

Fig. 6.3 Diets of three ground finch species on Daphne Major in three categories of seed size. Large seeds are *Tribulus cistoides*, medium seeds are *Opuntia echios*, and small seeds comprise more than 20 species. From Grant and Grant (2006a).

same amount but the effect upon population size was proportionally much greater, as only four females and nine males were left at the time breeding resumed in 2005. The population was almost extinct, apparently as a result of exhaustion of the standing crop of large seeds. Many birds of both species were found dead (Plate 25), and their stomachs were empty. There was little scope for character displacement in *G. magnirostris* because there are no seeds (fruits) larger or harder than *Tribulus* on the island. Although average beak size increased, both small-billed and large-billed birds survived (Fig. 4.5).

The principal alternatives to *Tribulus* for both species are the seeds of *Opuntia* cactus, but production in 2004 was unusually low. Not only were cactus seeds insufficient for the two granivore species to escape the dilemma of a diminishing supply of their preferred foods, they were insufficient for the cactus specialist *G. scandens* (Fig. 6.3; Plate 23), whose numbers, like those of *G. fortis*, fell lower than in any of the preceding 32 years: to 50. The only escape was available to the smallest, most *fuliginosa*-like members of the *G. fortis* population. These are known to feed like *G. fuliginosa* on very small seeds with little individual energy reward. It may be significant that two *G. fuliginosa* individuals were present on the island in 2004 and both survived to 2005.

Selection under Contrasting Conditions

Strong selection against large-beaked *G. fortis* in the presence of *G. magnirostris* in 2004 contrasts with the opposite in the absence of *G. magnirostris* in 1977 (Fig. 6.1, Table 6.1), that is, selection against small-beaked *G. fortis* (ch. 5). The selection events of 1977 and 2004 stand out against a background of relative morphological stability. Immediately prior to 2004 there was no unusual rainfall to cause a change in composition of the food supply, and no other unusual environmental factor such as temperature extremes or an invasion of predators, yet with the same amount of rain as in 1977 small finches survived at a high frequency in 2004 but survived at a low frequency in 1977. The conspicuous difference between these years was the number of *magnirostris*: 2–14 occasional visitors in 1977, 150 residents at the beginning of 2004. As a result the crash of the *G. fortis* population from 2003 to 2005 was more severe than the 1976–78 crash.

Evolution of Character Displacement

The shifts in average beak characters of the population were passed on to the next generation because they are highly heritable (Fig. 5.7). An evolutionary response is to be expected from strong directional selection, and can be predicted from the product of two quantities: the selection differential, which is a measure of the strength of selection, and the heritability, which is a measure of the amount of genetic variation (ch. 5). Offspring produced in 2005 by

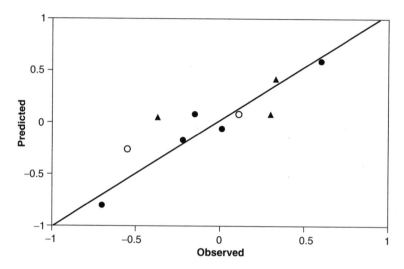

Fig. 6.4 Observed evolutionary reponses to natural selection on beak size (solid circles) and shape (open circles) in *G. fortis* and beak size in *G. scandens* (triangles) compared with predicted values. The reference line describes the relationship when observed = predicted. Predictions are the product of the selection differential and heritability in different selection episodes in standard deviation units. They are positive for size increases and negative for decreases. From Grant and Grant (2002a), with addition of the value associated with character displacement in *G. fortis* in 2005. This is the lowest point on the graph.

survivors of the long drought were close in average beak size to the predicted value (Fig. 6.4), and smaller than the parental generation's mean prior to selection by approximately 0.7 standard deviations. This fits the general pattern of a close correspondence between actual and predicted evolutionary change (Fig. 6.4).

In summary, average beak size of *G. fortis* became larger in the absence of *G. magnirostris* (1977); and in the presence of *G. magnirostris* it became smaller (2004), which is simultaneously displacement from *G. magnirostris* and release from *G. fuliginosa*. These evolutionary changes are more complex than those envisaged by Lack (1947). They provide direct support for the proposed morphological and ecological adjustments that competitor species make to each other during speciation. They add strength to the argument that competitive interactions are partly responsible for the pattern of regular beak differences we see between coexisting species (e.g., Fig. 1.3).

SUMMARY

A crucial stage in the cycle of speciation events is the establishment of sympatry by two previously separated populations derived from the same ancestor. Whether they interact competitively or not is determined by how different they are at that time. Interacting populations may be subject to natural selection that causes resource-exploiting traits (beaks) to diverge, and it results in diminished competition between them. This is referred to as character displacement. Character displacement in beak morphology is revealed by a comparison of populations of two closely related species in sympatry and allopatry. For example, *G. conirostris* is intermediate in beak form and body size between *G. magnirostris* and *G. scandens* in their absence (Española), and more like *G. scandens* in the presence of *G. magnirostris* (Genovesa). If, as appears to be the case, the Española population of *G. conirostris* is the older of the two, then evolutionary change on Genovesa is evidence of character displacement from the competitor *G. magnirostris*. Direct observation of the process of character displacement on Daphne Major strengthens the inference from an analysis of such patterns. *G. magnirostris* established a breeding population on the island in 1982, and gradually increased in numbers up to a maximum in 2003, a year of drought. By severely depleting the supply of *Tribulus* fruits during the drought, *G. magnirostris* caused an evolutionary shift in the *G. fortis* population in the direction of small beak size.

Reproductive Isolation

Geographically overlapping species of birds that are closely related,
when of similar appearance often possess markedly different songs.
(Huxley 1938, p. 257)

PRE-MATING BARRIER TO INTERBREEDING

THE CESSATION of interbreeding marks the end of the process of speciation. How does this come about? The answer is by an accumulation of differences between populations. Traits that function in the context of courtship diverge, just as traits that function in feeding diverge. According to the standard model, divergence does not occur *de novo* in sympatry, but begins in allopatry. Eventually the traits become so different that, when populations come together in sympatry, members of one do not recognize members of the other as potential mates. Differences in signaling traits and responses to them constitute a barrier to interbreeding. At this point pre-mating isolation of the two populations is complete. This chapter addresses the questions of what exactly the trait differences are that form the barrier to interbreeding between two recently established, coexisting populations, how those differences evolve in allopatry, and then how effective they are in sympatry.

A barrier preventing an exchange of genes may function before or after mating. Our concern in this chapter is solely with pre-mating isolation. Post-mating isolation arising from reduced viability or fertility of the offspring, or from their inability to acquire mates, will be considered in the next chapter.

FACTORS INVOLVED IN THE DISCRIMINATION BETWEEN SPECIES

All species of Darwin's finches have similar courtship behavior (Lack 1947, Ratcliffe 1981). Groups of related species in the same genus are identical in plumage; for example, in the ground finches (*Geospiza*) males are black and females are brown. Species differ conspicuously in beak size and beak shape, and also in song. Therefore we ask if individuals can discriminate between members of their own species and members of a closely related sympatric species on the basis of beak and song characteristics.

Beaks

David Lack (1945, 1947) proposed that the barrier to interbreeding was the finches' differing appearances, especially in beak size and shape, and these differences were acquired through natural selection in allopatry. If this is correct, the barrier to interbreeding arises as a by-product of ecological adaptation (Fisher 1930, Dobzhansky 1937, Schluter 2000) and is not built directly by natural (or sexual) selection.

The role of beak size and shape in courtship was first tested by Lack (1945, 1947) in preliminary experiments, and then by Laurene Ratcliffe on several islands with stuffed museum specimens of male or female ground finches (Ratcliffe and Grant 1983a). Contrasting specimens were set up at opposite ends of a rod attached to a tripod, one belonging to the tested bird's own population and the other belonging to a sympatric species (Plate 26). In several of these experiments the sizes of the specimens were similar and they differed almost exclusively in the beak. The design poses a question to the responding finch: can you tell the difference between the two individuals by appearance alone? The answer was yes. Responding birds attended more to their own population's specimen, which they courted or attacked depending on whether it was a male or a female, than to the other (Fig. 7.1).

Song

Discrimination on the basis of morphology is not the only possibility. Finch species also sing different songs, and Robert Bowman (1979, 1983) proposed that song differences prevented species from breeding with each other. The hypothesis of discrimination by song was tested on several islands with playback of tape-recorded song (Plate 26) to the same ground finch species used in the beak-discrimination experiments (Ratcliffe and Grant 1985). In the absence of visual cues, male finches did indeed discriminate between songs of their own and another sympatric species (Fig. 7.2). Male

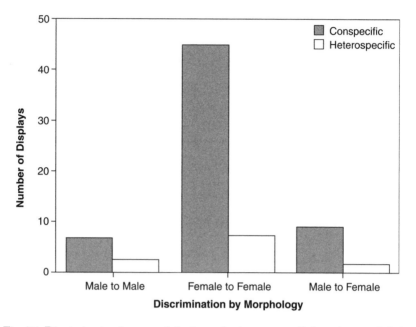

Fig. 7.1 Discrimination by ground finch species between stuffed specimens of their own local population (solid bars) and a different species (open bars) from the same island. Each bar represents the average time spent at each model by 8–16 birds on four islands: Daphne Major, Plaza Sur, Genovesa, and Pinta. The data from several experiments involving four species (*G. fortis, G. fuliginosa, G. scandens,* and *G. difficilis*) have been combined here. From Ratcliffe and Grant (1983a).

responses make sense if the stimuli come from sexual competitors. Therefore females, if given the opportunity to respond to playback by removing the males, would probably make the same discriminations. In the experiments females generally stayed away from the action, or were driven away by their mates. Studies of other species have circumvented this problem by using estradiol to elevate female responsiveness. For example, under these conditions female song sparrows, *Melospiza melodia,* responded to playback of tape-recorded songs and discriminated between local and heterotypic ones, favoring the former (Patten et al. 2004).

Thus individuals can discriminate between members of their own and a closely related sympatric species on the basis of song and beak morphology. The two cues differ in one fundamental respect: morphological traits vary genetically to a large degree (Fig. 5.7) whereas songs do not. Songs need

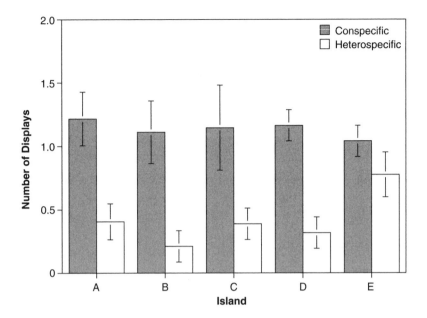

Fig. 7.2 Discrimination by male ground finches between local conspecific and heterospecific song in playback experiments on Plaza Sur (A), Pinta (B, C), and Daphne Major (D, E). Each bar represents an average value for approximately 10 birds tested with conspecific song (solid bars) and 10 different birds of the same population tested with heterospecific song (open bars). Vertical lines represent one standard error of the mean. The tested species and heterospecific song were *Geospiza fortis* tested with *G. fuliginosa* song (A), *G. fuliginosa* tested with *G. difficilis* song (B), *G. difficilis* tested with *G. fuliginosa* song (C), *G. fortis* tested with *G. scandens* song (D), and *G. scandens* tested with *G. fortis* song (E). Discrimination was shown by other response measures in E. From Ratcliffe and Grant (1985).

separate discussion because they are learned and not genetically encoded, and are therefore more susceptible to environmental influences.

LEARNING

Song learning has been demonstrated in captive rearing experiments in which isolated young Darwin's finches were exposed to tape-recorded songs of their own or a different species (Bowman 1979, 1983). Results may be summarized as follows. Song is learned through an imprinting-like process during a short sensitive period early in life, from either the father or another male; females do not sing. The time of learning, between approximately day 10 and day 40 after hatching, corresponds to the period from the last few days in the nest to the usual end of the fledglings' dependence on their parents. The father sings throughout this time.

Song is therefore a culturally transmitted trait, and as such is subject to non-genetic modification. Presumably if acoustic cues are learned early in life and used later in mate choice through sexual imprinting, so too are visual cues such as beak size and shape (Bowman 1983, Grant and Grant 1998a, 2002b, Grant et al. 2000). Learning to associate the two sets of cues begins in the nest, and continues during and possibly beyond the fledglings' period of dependence on their parents for food. The foundation of later mate discrimination, when mate preferences may be consolidated or modified by initial courtship experience (Bischoff and Clayton 1991), is built at this time when young birds are first exposed to the songs and appearances of other species. Learning what not to respond to accompanies learning what to respond to, and may be almost as important (Gill and Murray 1972, Lynch and Baker 1990, Grant and Grant 1996b). For example, early in life young male blackcaps (*S. atricapilla*) in Europe establish an association between visual (plumage) and acoustic (song) cues of their own species and of another species, the garden warbler (*Sylvia borin*); they retain the associations for eight months without heterospecific contact, and then use the information in the breeding season when discriminating against heterospecific individuals (Matyjasiak 2005).

Darwin's finches are not unusual in learning song. The phenomenon of song learning, as opposed to genetically determined song, is found in about half of all 27 orders of birds (ten Cate et al. 1993). The functional connection between song learning and speciation in birds has been made repeatedly (Payne 1973, Immelmann 1975, Clayton 1990, Irwin and Price 1999, ten Cate and Vos 1999, Payne et al. 2000).

SONG DIFFERENCES BETWEEN SPECIES

All sympatric species of Darwin's finches have their own characteristic songs (Bowman 1983). Songs are short, and individuals sing slightly different variants of the species song that we refer to as sub-types. Long-term studies of individually marked finches on Genovesa (Grant and Grant 1989) and Daphne (Gibbs 1990, Grant and Grant 1996b) show that male ground finches, with rare exceptions, sing only one song, usually the same type as their father's (Fig. 7.3), and once learned the song is retained for life (Fig. 7.4). On Daphne Major the cactus finch *G. scandens* (Fig. 7.5) sing a very different song from the medium ground finch *G. fortis*, and the two species ignore each other's tape-recorded songs when played to them (Ratcliffe and Grant 1985). The learned differences constitute a barrier to interbreeding. Parallel examples with different species occur on Genovesa. Just how great the differences must be to constitute a barrier is unknown: perhaps fifteen percent along a salient axis of variation, as in beak size (ch. 6). The axis or axes have yet to be identified.

Therefore reproductive isolation of Darwin's finches is explained by a theory of discrimination by song in association with morphology that is learned in a sexual imprinting-like process early in life and usually from parents (Grant and Grant 2002b, 2002d). Finches learn the two cues interdependently, and possibly in a mutually reinforcing way. How they weigh the two cues when attempting to discriminate among individuals of their own species, and between them and another finch species, is difficult to gauge (Grant et al. 2000, Grant and Grant 2002b), in part because we still do not know if females choose on the base of performance-related features of song (trill rate and frequency range; Podos et al. 2004a, 2004b), phoneme structure and temporal patterning, duration and uniformity of inter-note intervals or of the notes themselves (Grant and Grant 2002b), frequencies at which sound energy is maximal, unmeasured tonal features, or some complex combination.

These remarks concerning early learning apply to both song production (males) and song recognition (both sexes). A similar situation exists in the well-studied zebra finch (*Taeniopygia guttata*). Females do not sing, but preference tests with sibling groups have shown that both sexes have equal preference for their father's song over an unfamiliar male song (Riebel 2000, Riebel et al. 2002). They also imprint on colorful plumage features and beak colors of their parents (Clayton 1988, ten Cate et al. 2006).

SONG DIVERGENCE IN ALLOPATRY

Songs diverge in geographical isolation. Song differences are most pronounced among populations of the sharp-beaked ground finch (*Geospiza difficilis*), which

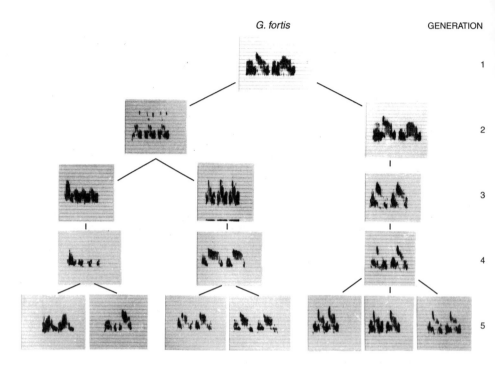

Fig. 7.3 A pedigree of songs sung by *Geospiza fortis* males on Daphne Major, to illustrate the fact that most but not all sons sing songs identical to their fathers' songs. Frequency from 0 to 8 kHz on the vertical axis is plotted against time in seconds. From Grant and Grant (1996b).

also differ in beak size and feeding habits (ch. 5). On the three high islands of Fernandina, Santiago, and Pinta (Figs. 1.1 and 5.3), these finches occupy structurally complex and occasionally dense *Zanthoxylum* forests (Plates 7 and 12), which impedes long-distance transmission of sound. The songs are complex in note form, temporal pattern, and frequency modulation and have high frequencies (Fig. 7.6). In contrast, on the low islands of Genovesa, Wolf, and Darwin the habitat is structurally more open and simple, vegetation does not obstruct long-distance transmission of sound to the same extent, but seabirds call loudly in the low frequencies. Finch songs are short and structurally simple. They comprise a repetition of simple notes, or a narrow-band buzz, without strong frequency modulation but with initial "alerting" notes (Fig. 7.6). At some point in their history populations of this species diverged

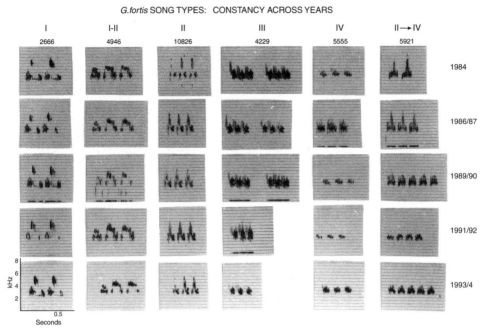

Fig. 7.4 Lifetime constancy of the single song sung by *Geospiza fortis* males on Daphne Major. From Grant and Grant (1996b).

in song characteristics, and did so in a way that suggests habitat was an important influence.

Adaptation to Habitat

Properties of song could change under sexual selection in structurally different habitats on different islands (Bowman 1983, Price 1998). Songs could change because some are more attractive than others to females (inter-sexual selection) or because some are more effective at repelling other males that are sexual competitors (intra-sexual selection). A more efficient transmission of sound in the habitat occupied should result in higher mating success from both causes, all else being equal. Song frequencies that birds emphasize vary among islands, and generally correspond with maximal sound energy transmission through the vegetation (Bowman 1983). This correspondence supports the suggestion of adaptive change in song features, principally in pitch and frequency modulation (Bowman 1983).

83

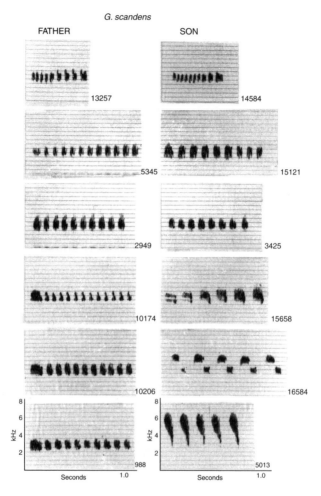

Fig. 7.5 Similarity of *Geospiza scandens* songs produced by father and son on Daphne Major (upper three) with some variation (sons, lower three). From Grant and Grant (1996b).

Change of Songs as a Consequence of Morphological Divergence

Properties of song could also change simply as a consequence of a change in the size of the bird. Bowman (1983) showed that the median frequency of finch song decreased among species as their body sizes, and the sizes of their internal resonating chambers, increased. There is very large variation among individuals of the same species, however.

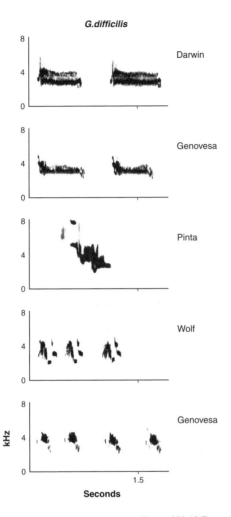

Fig. 7.6 Advertising songs of *Geospiza difficilis* on Pinta, Wolf, Darwin, and Genovesa. From Grant et al. (2000).

Two other song features, note repetition rate and range of frequencies, vary systematically with beak and head size; the larger the beak, the slower are the notes produced and the smaller is the range of frequencies (Podos 2001, Podos et al. 2004a, Huber and Podos 2006). The association is statistically strong and has a biomechanical explanation: large species cannot repeatedly open (widely) and close their beaks as fast as some small species can (Podos et al. 2004b). However, there are exceptions to the trend (Huber and Podos 2006): for example, the note repetition rate of *G. difficilis* is faster in the songs on Wolf than on Genovesa, where beak size is much smaller, which is opposite to what is expected. The sister species *G. fortis* and *G. fuliginosa* on Santa Cruz do not differ discretely in these song features (Grant and Grant 2002d), and some songs are more striking for their similarities than for their differences (Fig. 7.7; other examples are illustrated in Bowman 1983). In several cases a species is known to occasionally sing an almost identical version of another species' song, and the other species responds to it (Bowman 1983, Grant and Grant 1996b; see also next chapter). Aside from these and other exceptions (Slabbekoorn and Smith 2000, Grant and Grant 2002d, Seddon 2005), a possible consequence of adaptive change in beak morphology in allopatry is an alteration of vocal characteristics sufficient to make a male sound strange to members of a related population (Podos and Nowicki 2004).

Since beak and body size are generally strongly correlated, they are likely to be jointly involved in determining song characteristics (Grant and Grant 2006b). Separately or together these explanations are consistent with the thesis of Fisher (1930) and Dobzhansky (1937) that barriers to interbreeding arise as a by-product of adaptive divergence (Barton 2001). In the first they are a product of exploiting a particular habitat, and in the second they are a by-product of adaptive change in morphology. The common denominator is feeding ecology.

The Role of Chance

An ecological theory of speciation (Schluter 1998, 2000) is not sufficient to explain the divergence of songs in allopatry. Chance plays a significant role too.

Minor differences between individuals in the elements or motifs (phonemes) of learned song are believed to come about through an accumulation of copying errors. Most novelties are likely to be lost from the population by chance, but by chance some may increase and drift to high frequency. An increase could also be actively propelled by female choice. Given enough time, the differences between populations could become large enough to function as a barrier to interbreeding. Environmental differences need play no role in the divergence (Price 2007).

Not only are novel variants generated by chance, through mis-copying, but there is the possibility of random loss of elements of a song, for example in the founding of a new population by a few individuals (ch. 4). This is

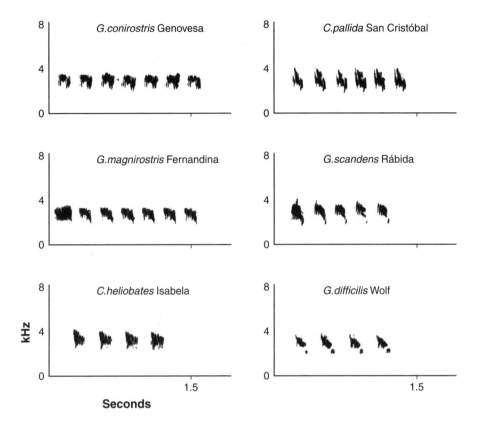

Fig. 7.7 Song similarities among species. Some songs can be distinguished by subtle features that are not captured by these sonagrams. The song of *G. magnirostris* recorded on Fernandina (in 1997) sounded to our ears like *G. conirostris* or *C. pallidus*, neither of which is resident on Fernandina!

especially likely if the founders are young birds capable of singing only an incomplete rendering of song (Thielcke 1973). Warbler finches provide an entirely hypothetical example. The short song motifs sung by members of the *olivacea* lineage could have been derived this way from the longer songs of members of the *fusca* lineage. Novel variants may also increase through chance, as occurred on Daphne Major island when a single immigrant male *G. magnirostris* introduced a novel song variant when the population was small (ch. 4). This male was a productive breeder and its song type rapidly increased in frequency of occurrence, becoming the most prevalent sub-type in the population (Grant et al. 2001).

On Wolf and Darwin, two neighboring islands in the northwest of the archipelago, sharp-beaked ground finches (*G. difficilis*) are almost identical in morphology and their habitats are almost identical structurally, yet their songs are surprisingly and strikingly different (Fig. 7.6). If the two islands were colonized separately by members of an ancestral population that, like the phylogenetically older warbler finch, had both repeated notes and buzzes in the total repertoire, there could have been an element of chance in which part of the ancestral song was lost and which part was retained in each of the small populations, either in the initial founding of each population or later during times of very low numbers (Grant et al. 2000).

In conclusion, a premating barrier to interbreeding based on song arises in allopatric populations, partly as a result of adaptation to the environment and partly as a result of chance.

SIMULATING SECONDARY CONTACT

Song differences in geographical isolation imply incipient speciation, providing the differences can be discriminated by finches and are not solely detected by us. The potential for discrimination cannot be tested by putting members of separate populations together in the Galápagos National Park, but it can be tested by using tape-recorded song. Secondary contact of previously separated populations is simulated by playing back to birds on island A songs that were recorded on island B. Such experiments have demonstrated relatively weak discrimination between songs from a high island (Pinta) and a low island (Genovesa) form of *G. difficilis* (Ratcliffe and Grant 1985).

Even within the group of three lowland populations there is discriminable variation among populations. Birds on Genovesa sing two types of song, either a repeated-note song or a buzzy song, and very rarely both of them (Grant et al. 2000). Males respond equally to both song types. Birds on Darwin sing only the buzzy song, which is almost identical to the buzzy song of Genovesa birds, and on Wolf they sing only a repeated-note song (Fig. 7.6). Playback experiments on Genovesa showed that regardless of the type of song they themselves sing (repeated or buzzy), males respond to the Darwin song almost as strongly as if it was the same as their own (Grant and Grant 2002b). On the other hand, birds of both song types discriminated strongly against songs of the Wolf birds; all but one of 12 tested birds ignored the Wolf songs. Possibly

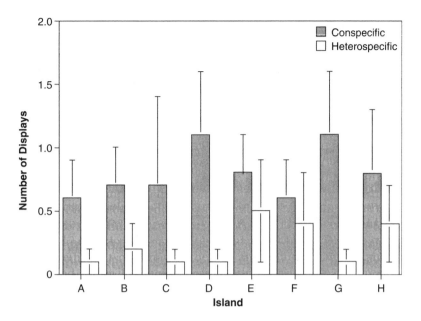

Fig. 7.8 Discrimination by ground finch species between stuffed specimens of their own local population (solid bars) and a morphologically similar species from a different island (open bars), an immigrant. Each bar represents average time spent at each model by approximately 10 birds tested in one location; vertical lines indicate one standard error. Identities of island, tested species, and the immigrant are as follows: A Daphne *G. fortis* with Santa Cruz *G. fuliginosa*; B Española *G. fuliginosa* with Daphne *G. fortis*; C Española *G. fuliginosa* with Santa Cruz *G. fortis*; D Plaza Sur *G. fortis* with Santa Cruz *G. fuliginosa*; E Plaza Sur *G. fuliginosa* with Santa Cruz *G. fortis*; F Plaza Sur *G. scandens* with Genovesa *G. conirostris*; G Plaza Sur *G. scandens* with Genovesa *G. difficilis*; H Genovesa *G. difficilis* with Santa Cruz *G. scandens*. From Ratcliffe and Grant (1983b).

they responded to the difference in temporal pattern: five evenly spaced notes (Wolf) versus similar notes sung more slowly and in pairs (Genovesa).

These experimental results demonstrate incipient speciation to varying degrees as a result of apparently small song differences acquired allopatrically. Birds flying from one island to another would encounter relatives that sang a different song, the degree of difference depending on the particular island. The difference to their ears could be so large that they would ignore the alien song.

The same kinds of experiments were performed with stuffed specimens (Fig. 7.8), but in the absence of song (Ratcliffe and Grant 1983b). Results of

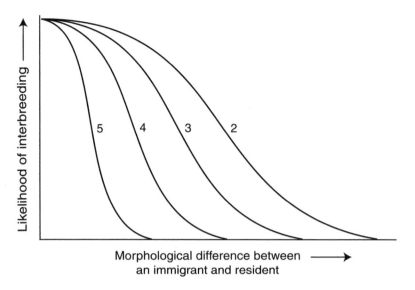

Fig. 7.9 The likelihood of a resident species breeding with an immigrant declines in proportion to the difference in their morphology; the smaller the morphological difference, the greater is the likelihood of interbreeding. The decline varies with the size of the island and the number of related species it supports. On an island with five species (5) the decline is steep, whereas on an island with few species (2) it is much more gradual: interbreeding is more likely on a species-poor island than on a species-rich island, according to the theory of learning (see text). From Grant (1999).

morphological experiments were similar to those using song alone, and like them they imply incipient speciation. Not surprisingly, discrimination was strongest when the "immigrant" was markedly different in morphology from the resident. For a given morphological difference, discrimination was stronger on islands with several species of ground finches than on islands with fewer species (Fig. 7.9). This is consistent with the idea that finches learn what not to respond to from their early experience, and has some interesting implications. It implies, for example, that immigrants and residents are least likely to interbreed if the immigrants come to a large island with a large community of finches, from an island with a similarly large community of finches. The opposite is to be expected on small islands. Yet small islands, especially well isolated peripheral ones, are sources of evolutionary novelty because their environments are unusual and populations are small (Petren et al. 2005). Therefore speciation might be initiated by divergence in song and beaks most often in small peripheral islands (allopatry) and continue on large islands (sympatry) where residents will not interbreed with immigrants from the small islands.

SUMMARY

Darwin's finches have similar courtship behavior, and groups of related species in the same genus are identical in plumage. They differ conspicuously in beak size and beak shape as well as in song. Field experiments show that individuals are capable of discriminating between their own and another species on the basis of song and morphology. Both morphological and vocal differences constitute a pre-mating barrier to interbreeding, and the barrier arises in allopatry. Unlike beak traits, which are genetically inherited, song is learned and is therefore a culturally transmitted trait. Learning takes place during a short sensitive period early in life in an imprinting-like process. The visual cues of beak size and shape are presumably learned at the same time. These cues diverge adaptively (ch. 5). Associated with adaptive morphological change, properties of song may change simply as a passive consequence of a change in body size; the larger the bird, the lower the pitch. Evidence is mixed for the theory that song properties change as a passive consequence of a change in beak size and shape; the larger the beaks, the slower the note-repetition rate and the smaller the range of frequencies. Vocal variation is believed to be subject to sexual selection, resulting in more efficient transmission of sound in the habitat occupied. These facts are consistent with Fisher's and Dobzhansky's thesis that barriers to interbreeding arise as a by-product of adaptive divergence. Some change, however, is non-adaptive. Songs diverge through the accumulation of different copying errors (cultural mutations) in allopatric populations. Chance may play an additional role in the random loss of different elements of a song in the founding of a new population or later when numbers are low.

CHAPTER EIGHT

Hybridization

> [T]he extraordinary variants that crop up in a series [of
> museum specimens] give an impression of a process of change and
> experiment going on.
> *(Swarth 1934, p. 231)*

INTRODUCTION

AS DESCRIBED in the previous chapter, beak morphology and song establish a barrier to interbreeding, and the barrier evolves in allopatry. Is the barrier fully formed at the time of secondary contact? If it is not, then some of its elements could be subject to selection, causing divergence, consolidation (reinforcement) of the barrier, and complete cessation of interbreeding. Sexual selection on factors affecting mate choice is to be expected if there is a post-mating disadvantage to interbreeding—that is, if offspring produced by interbreeding, the hybrids, are relatively unfit, because they do not survive, acquire mates, or reproduce as well as the offspring produced by matings within each population. In this chapter we examine the incidence and causes of hybridization and the fitness of hybrids, before discussing the likelihood that the barrier to interbreeding is strengthened (reinforced) in sympatry.

HYBRIDIZATION

Darwin's finch species hybridize, albeit rarely, and hybrids survive to breed. This shows that the barrier to interbreeding is not complete in all cases; it leaks. Permeability of the barrier further demonstrates that speciation is not completed in allopatry, as do the experiments described in the previous chapter.

The principal evidence for hybridization comes from observations of pairing patterns of banded birds. The long-term studies of banded finches on Genovesa and Daphne showed that interbreeding of ground finch species occurs in most years at a frequency of one percent or a little higher; in other words, rarely but persistently. On Genovesa *G. conirostris* hybridizes both with the larger species *G. magnirostris* (Plate 27) and the smaller species *G. difficilis* (Grant and Grant 1989). On Daphne Major (Plate 28) *G. fortis* hybridizes with the larger *G. scandens* and with occasional immigrants of the smaller species *G. fuliginosa* (Grant and Grant 1992, Grant 1993). Because paternity can be wrongly identified by observations alone, microsatellite DNA markers were needed to confirm that hybridization was correctly inferred when we identified males and females, the "social parents," by their feeding of young at a nest (Grant et al. 2004). This was necessary in view of evidence from studies elsewhere that apparently hybridizing birds may produce non-hybrid offspring through extrapair copulations with conspecific mates (Slagsvold et al. 2002, Veen et al. 2001, Reudink et al. 2005).

Hybridization is not restricted to the relatively small islands of Daphne Major and Genovesa, or to ground finches. A pairing of a small tree finch (*Camarhynchus parvulus*) and a warbler finch (*Certhidea olivacea*) on Santa Cruz is thought to have led to backcrossing to *C. parvulus* (Bowman 1983, Grant 1999). A few specimens in museum collections are difficult to classify because they are morphologically intermediate between sympatric species (Fig. 1.3). They could be hybrids. There are two specimens from Floreana that are intermediate between *C. parvulus* and the *fusca* lineage of *C. olivacea* (Swarth 1931, Lack 1945, 1947). Specimens from Santa Cruz are intermediate between the woodpecker finch (*Camarhynchus* [*Cactospiza*] *pallidus*) and the warbler finch (*Certhidea olivacea*), and between the small tree finch (*Camarhynchus parvulus*) and the warbler finch (*Certhidea olivacea*). If our current genetic analysis (with Ken Petren) of material from museum specimens confirms their hybrid identity they would represent successful interbreeding of species that shared a common ancestor ~2 MYA.

WHY HYBRIDIZATION OCCURS

It is well known that introgressive hybridization occurs in plants and animals when the environment is disturbed, particularly by humans (Anderson 1948,

Stebbins 1959, Arnold 1997, 2006, Seehausen et al. 1997, Taylor et al. 2006). Habitat disturbance may bring together previously separated populations, or weaken the normal barriers to gene exchange by increasing opportunities for interbreeding, leading to fusion of the interbreeding species. It may also create new habitats that are favorable for the hybrids and backcrosses, leading to the establishment of a zone of hybrids between the habitats of the interbreeding species. Hybridization of Darwin's finches is unusual though not unique in occurring in an entirely natural, undisturbed environment. It is not confined to a geographically restricted hybrid zone, unlike many other examples elsewhere (Harrison 1993, Rowher et al. 2001, Secondi et al. 2003, Bronson et al. 2005), nor is it restricted to sister species.

In general hybridization occurs when species have similar courtship signals and responses. What are the most important signals? Species with similar beak and body sizes hybridize, therefore it might be expected that the interbreeding individuals of each species are particulary similar in morphology. For example, the largest of the small ground finches (*G. fuliginosa*) might breed with the smallest of the medium ground finches (*G. fortis*) because they are more similar to them than to the smallest members of their own population (Fig. 1.3). However, despite this striking fact, hybridization of most similar individuals is not the predominant pattern on Daphne Major (Boag and Grant 1984b). This shows that similarity in beaks, and in any vocal charcteristics that might be correlated with beaks, does not primarily govern the choice of mates in interbreeding species. Instead, the evidence points to misimprinting on the song of another species as the cause of interbreeding.

An imprinting-like process (ch. 7), involving learning, is vulnerable to perturbation if the young bird hears the song of another species rather than a conspecific song during the short sensitive period. The bird then misimprints on heterospecific song. This occurs rarely, in less than one percent of young, and under a variety of idiosyncratic circumstances: following the death of the father, for example, or when a loud male of one species (*G. magnirostris*) repeatedly drives away the male of another species (*G. fortis*) from its unusually close nest, and persistently sings (Fig. 8.1). In one instance on Daphne Major a pair of *G. scandens* appropriated the nest of a pair of *G. fortis*, and did not remove one egg. The male that hatched from the egg was raised by the pair of *G. scandens*. It learned, and later sang, the song of its foster father (Grant and Grant 1996b, 1997a, 1998a).

MISIMPRINTING

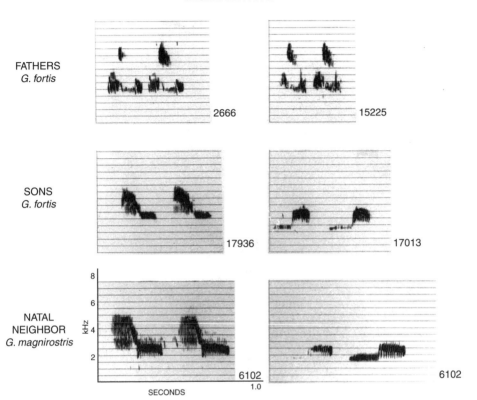

Fig. 8.1 Miscopying of *G. magnirostris* songs by *G. fortis* on Daphne Major as a result of an imprinting-like process. The two *G. fortis* sons copied two songs sung by a neighboring *G. magnirostris* male and not the songs sung by their respective fathers. Male 15225 was the son of 2666: a rare case of a son occupying a territory next to his father. The *G. magnirostris* male is also a rare example of an individual singing two songs; by chance his songs rapidly replaced two other sub-types in the population (ch. 7). In neither case did the misimprinting lead to hybridization. From Grant and Grant (1998a).

We cannot identify the specific causes of every individual interbreeding event by direct observation. For example, it is a mystery why one daughter of a pair of *G. fortis* on Daphne Major paired with a *G. scandens* male, and a sister raised in the same nest paired in the same year with a *G. fuliginosa* male (Grant and Grant 1997a)!

Misimprinted birds sometimes, but not always, mate with a member of the species on which they have imprinted: they hybridize. We recorded the songs of both the father and the mate of 482 female finches (*G. fortis, G. fuliginosa,* and *G. scandens*) on Daphne Major, and of these, 16 hybridized, all mating according to paternal song type and not according to parental beak morphology. In 12 cases females mated with misimprinted males of another species, and in the remaining four cases females were the offspring of a misimprinted father (Grant and Grant 1996b).

Thus misimprinting on song can lead to hybridization. It is significant for another reason. Faithful copying of the song of one species by another shows that some morphological constraints from beak size on sound production (ch. 7) can be overcome, or at least minimized and functionally neutralized.

When Hybridization Does Not Occur

When species become very different in size they do not interbreed, or do so extremely rarely. On Genovesa we never observed interbreeding between the largest species, *G. magnirostris*, and the smallest species, *G. difficilis*, and on Daphne Major the largest species, *G. magnirostris*, has not bred with any of the other species. Differences in size are confounded with the length of time they have been separate species, but events on Daphne Major show that size differences are the more important factor.

G. magnirostris and *G. fortis* are close relatives if not actual sister species (Fig. 2.1). A total of six *G. fortis* males are known to have misimprinted on *G. magnirostris* song on Daphne Major (e.g., Fig. 8.1), but none of them paired with a *G. magnirostris* female. The size difference appears to be the main barrier to interbreeding. *G. magnirostris* (> 30 g) is appreciably larger than *G. fortis* (~18 g) on this island, whereas *G. scandens* (~22 g), which has bred with *G. fortis*, even though it is not its closest relative, is much closer in size to it. Lack of interbreeding may signify that a male is not recognized as a potential mate by a female because at least one of the two cues she has learned (song and morphology) is too different. This possibility could be tested with *G. magnirostris* females. Observations suggest an additional possibility: prevention of interbreeding by male harassment. All six misimprinted *G. fortis* males were repeatedly chased by territorial *G. magnirostris* males, even though in some cases their songs were slightly higher in average frequency than the *G. magnirostris* songs they had copied. This demonstrates that song was a potent stimulus. One of the *G. fortis* males paired, with a *G. fortis* female, only after becoming almost silent!

HYBRID FITNESS

Fitness varies. From 1976 to 1982 none of the hybrids on Daphne Major survived long enough to breed. Many offspring produced by *G. fortis* and *G. scandens* also died at this time, although first-generation (F_1) hybrids produced by *G. fortis* × *G. fuliginosa* pairs fared significantly worse than offspring of *G. fortis* pairs: they were relatively and absolutely unfit (Grant 1993, Grant and Grant 1993). At the time we thought that a possible explanation for the poor survival of hybrids was genetic incompatibility of the genomes of their parents. However, an alternative explanation was that hybrids, being intermediate in beak morphology between the parental species, lacked a sufficient supply of seeds of the appropriate size to survive the dry season. This alternative is supported by evidence of feeding behavior. The large and hard *Tribulus* mericarps on which *G. fortis* were surviving during this period were too large for hybrids to crack, and even though some attempted the task none were seen to be successful. Moreover, although some hybrids can exploit *Opuntia* seeds (Fig. 8.2), the main dry season food of *G. scandens*, they do so significantly less efficiently than *G. scandens* (Grant and Grant 1996c).

The 1983 El Niño event perfoundly changed the ecological conditions (ch. 5), and the seed bank became dominated by small and soft seeds from 22 species of plants (Fig. 8.3). Under these altered conditions survival of hybrids with their intermediate bill sizes and shapes was high. This applies to F_1 hybrids of both *G. fortis* × *G. fuliginosa* pairs (Fig. 8.3) and *G. fortis* × *G. scandens* pairs (Fig. 8.4), and to the backcrosses, that is the offspring of F_1 hybrids mated to parental species. A comparison of pure species (*G. fortis* and *G. scandens*), F_1 hybrids, and backcrosses of the 1983, 1987, and 1991 cohorts, chosen for the years of maximum fledgling production, revealed that hybrids and backcrosses survived as well as, if not slightly better than, the pure species hatched at the same time (Fig. 8.4) and experiencing the same environmental conditions at the same age (Grant and Grant 1998a). Furthermore they had no difficulty in acquiring mates, and they reproduced as well as the parental species, there being no difference in number of eggs, nestlings, or fledglings produced (Grant and Grant 1992). Thus in neither survival not reproduction were hybrids and backcrosses at a fitness disadvantage. This demonstrates that *G. fortis* are genetically compatible with *G. fuliginosa* and *G. scandens*, and that

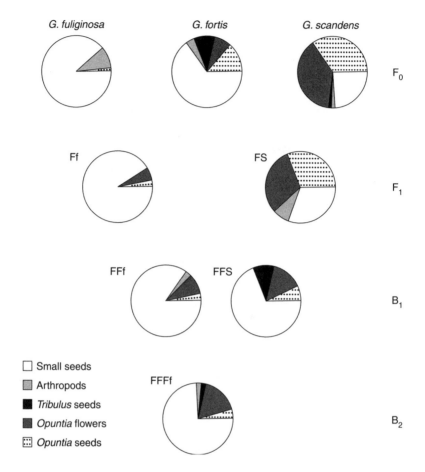

Fig. 8.2 Diets of ground finch species (F_0), F_1 hybrids, and two generations of backcrosses (B_1 and B_2) on Daphne Major. Symbols refer to the proportions of genes from *G. fuliginosa* (f), *G. fortis* (F), and *G. scandens* (S) in the F_1 hybrids; the proportion of genes in the backcrosses is underrepresented by the abbreviated symbols. From Grant and Grant (1996c).

the relative fitness of hybrids is dependent on external ecological conditions (Grant and Grant 1992). The enhanced heterozygosity of the hybrids may have contributed to their high fitness, alleviating a small deleterious effect of inbreeding (Grant et al. 2003).

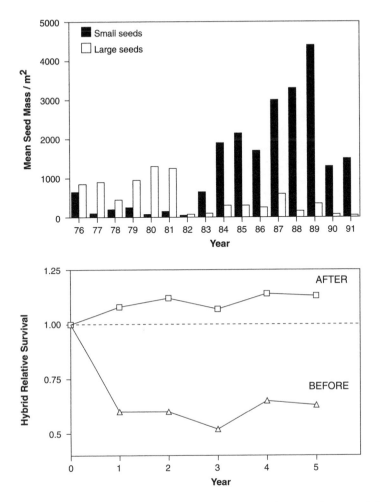

Fig. 8.3 Change in the fate of hybrids on Daphne Major after abundant rain fell in 1982–83 (Fig. 5.5). Biomass (mg) of small seeds (black bars) became much greater than the biomass of large seeds (white bars, upper diagram), and the relative fitness of *G. fortis* × *G. fuliginosa* hybrids greatly improved (lower). Relative fitness is expressed as hybrid survival compared with *G. fortis* survival: < 1.0 or > 1.0. Before the El Niño event hybrids were relatively unfit (1976–81 cohorts, BEFORE; < 1.0) but they were marginally fitter than *G. fortis* after the event (1983–87 cohorts, AFTER; > 1.0). From Grant and Grant (1993).

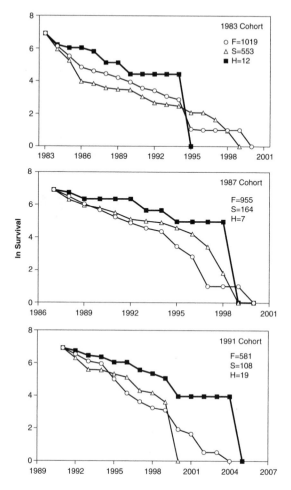

Fig. 8.4 Fitness of hybrids and backcrosses in relation to the parental species on Daphne Major. Survival is on a natural log scale, with initial numbers indicated in each box scaled to 1000. Hybrids and backcrosses (H) are shown with solid squares, *G. fortis* (F) with circles, and *G. scandens* (S) with triangles.

INTROGRESSION ON DAPHNE MAJOR

First-generation hybrids are too rare to breed with each other and produce an F_2 generation; we know of only two exceptions. Instead they backcross to one parental species or the other (Plate 28) except to the rare *G. fuliginosa*. High survival rates of hybrids and backcrosses from 1983 to 2006 and frequent

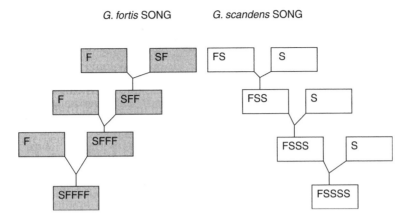

Fig. 8.5 Hybridization and backcrossing of *G. fortis* (F) and *G. scandens* (S) on Daphne Major proceeds according to paternal song type and leads to introgression of alleles. FS refers to an F_1 hybrid that sings a *G. scandens* song. When this happens it breeds with a *G. scandens* and produces first-generation backcross offspring (FSS). The same labeling convention is used for an F_1 hybrid that sings a *G. fortis* song (SF), and for later-generation backcrosses (SFF, etc). The convention, as well as the mating pattern itself, applies equally to females.

breeding resulted in introgression of genes from *G. fuliginosa* to *G. fortis*, and between *G. fortis* and *G. scandens*. Hybrid and backcross individuals are interesting because they broadcast potentially conflicting signals of their identity: the song of one interbreeding species or the other, and morphology that is intermediate between the two. The pattern of mating reveals the greater importance of song in mate choice (Grant and Grant 1998a). Like the parental species themselves that occasionally hybridize, if an F_1 hybrid is a male it sings the same single song as its father's and mates with a female whose father sang that song type (Grant and Grant 1997b). If it is a female it chooses a mate that sings the same song as its father's. In other words, hybrids backcross according to song type (Fig. 8.5).

Morphology appears to play a subordinate role in the choice of mates, as illustrated with the following two examples. The first is the pairing pattern of the most *fortis*-like hybrid produced by a male *G. scandens* and a female *G. fortis*. The hybrid male, who sang the same *scandens* song as his *G. scandens* father, first paired with a female *G. fortis*, and then with a female *G. scandens*. Later in

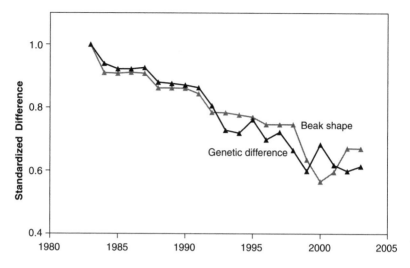

Fig. 8.6 Genetic and morphological convergence of *G. fortis* and *G. scandens* on Daphne Major as a result of introgressive hybridization. Standardization of differences between the species in beak shape and microsatellite alleles was achieved by giving a value of 1.0 to the differences in 1982. From Grant et al. (2004).

life he paired with a female of each species again. Microsatellite markers confirmed his paternity of the offspring produced from each pairing. In the second example, four daughters of a misimprinted *G. scandens* father paired, as expected, with *G. fortis* males singing typical *fortis* songs. The four males were a non-random selection of available mates in morphology; they were more *scandens*-like than the average in large size and relatively pointed beaks (Grant and Grant 1997b).

Although introgressive gene flow is bidirectional between *G. fortis* and *G. scandens*, from 1990 onwards it was three times greater from *G. fortis* to *G. scandens* than vice versa. At this time there were more *G. scandens* males than females, and all mixed pairs were formed by *G. scandens* males and *G. fortis* females (Grant and Grant 2002c). The rate of hybridization continued to be low (< 2%) and survival of hybrids and backcrosses remained high, with the result that by 2003 approximately 30% of *G. scandens* individuals contained some *G. fortis* alleles. Populations of the two species had become more similar to each other both genetically and morphologically (Grant et al. 2004) (Fig. 8.6), especially in beak shape (Fig. 5.9); the morphological tract in which they had previously differed the most.

INTROGRESSION IN THE ARCHIPELAGO

Hybridization has not been studied elsewhere in the archipelago apart from Genovesa (Grant and Grant 1989), but its occurrence can be inferred from morphological patterns (Fig. 1.3), as well as from genetic information according to the following reasoning. If selectively neutral genes have been exchanged at a low rate between hybridizing species, then two species on an island will be more similar to each other genetically than either will be to populations of the other species on another island. A study using microsatellite markers (Grant et al. 2005a) found a very strong tendency in the expected direction, both among ground finch species and among tree finch species (Fig. 8.7). The most extreme example is *G. fortis* and *G. fuliginosa*, two genetically and morphologically distinct species. On every one of eight islands where the two species occur together, each is more similar to the other species on the same island than it is to the other species on all other islands.

REINFORCEMENT

Reinforcement is the enhancement of differences in reproductive characters in sympatry as a result of selection against hybrids. Dobzhansky (1937) proposed that large differences between coexisting species arise in the terminal phase of speciation by this means: the most similar members of the two populations are selected against because they occasionally interbreed, and when they do they produce relatively unfit hybrid offspring (Liou and Price 1994, Hostert 1997). There is evidence for reinforcement in birds (Sætre et al. 1997) as well as a variety of other organisms (e.g., Coyne and Orr 1989, 1997, Marshall and Cooley 2000, Geyer and Palumbi 2003, Nosil et al. 2003, Pfennig 2003, Hoskin et al. 2005, Lukhtanov et al. 2005, Peterson et al. 2005).

Reproductive isolation in Darwin's finches is apparently solely a phenomenon of divergence in courtship signals and responses. There is no evidence of developmental problems arising from genetic incompatibilities after offspring (zygotes) have been formed (Grant and Grant 1997c). Post-zygotic

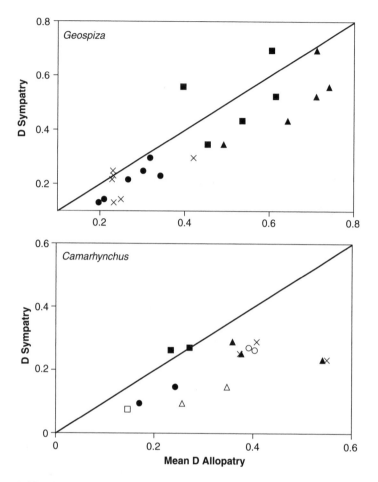

Fig. 8.7 The footprint of introgressive hybridization between closely related pairs of ground finches (upper) and tree finches (lower) in the Galápagos archipelago. A predominance of values below the lines of equality shows that genetic differences (D) between species are generally smaller in sympatry than in allopatry. Different symbols refer to different species. These are identified in Grant et al. (2005a).

incompatibilities that cause inviability or infertility begin to evolve after populations have been separated for 2–3 MY, as Price and Bouvier (2002) showed in a broad survey of birds. Most if not all Darwin's finches diversified in less than this time, therefore absence of genetic incompatibilities is not surprising.

Despite the absence of genetic incompatibilities, hybrids and backcrosses may nonetheless be at an ecological disadvantage under conditions of scarcity of the particular foods they are best equipped to exploit. Selection against interbreeding members of two populations could in principle operate then, strengthening the barrier against interbreeding (Huxley 1938). The barrier would be relatively weak because individuals of two species that interbreed are generally not particularly similar to each other in morphology (Boag and Grant 1984a, Grant and Grant 1997a), nor are they in features of their species-specific songs (Grant and Grant 1997a). Conversely, those that do not interbreed differ little if at all from those that do interbreed. The strength of selection against hybrids and backcrosses would also be a function of inter-breeding frequency, which is rare, and the length of time unfavorable ecological conditions persist. Unfavorable, generally dry, conditions for hybrid survival appear to be associated with relatively low sea surface temperatures, the so-called anchovy years, and favorable conditions occur during wetter periods of relatively high temperatures, the "sardine years" (Chavez et al. 2003). Sea sur-face temperature conditions alternate at approximately 25-year intervals. Thus the net strength of selection against hybrids should be computed after 50 years, and may be negligible on that time scale.

If reinforcement of song differences has occurred it can be detected, like character displacement, by comparing differences between a pair of coexisting species with allopatric populations that might have given rise to them. Laurene Ratcliffe (1981) did just this with a multivariate analysis of seven song charac-teristics of the ground finch species on many islands, and found scarcely any evidence for divergence in sympatry.

Reproductive Character Displacement

Beak size is a trait that functions daily as an instrument for gathering food, but in the breeding season it also provides a signal of identity to finches seeking a mate. Character displacement of beaks under ecological selection pressures in the non-breeding dry season (ch. 6) might diminish the chances of interbreed-ing. If so, it may be described as reproductive character displacement as well. Effects of character displacement on mate choice, however, are simply a con-sequence and not a driving force of the divergence.

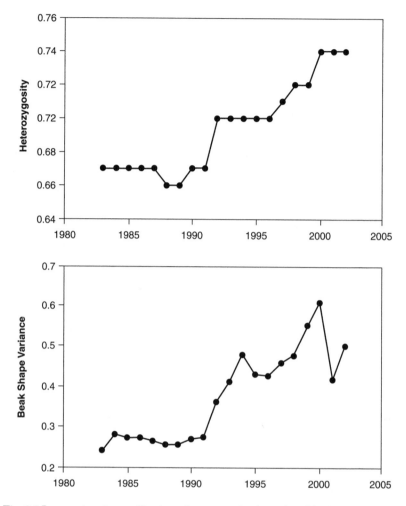

Fig. 8.8 Increase in microsatellite locus heterozygosity (upper) and beak shape variance (lower) in *G. scandens* on Daphne Major as a result of introgressive hybridization with *G. fortis*. From Grant et al. (2004).

EVOLUTIONARY SIGNIFICANCE OF INTROGRESSION

Introgression could have been important in facilitating rapid evolution when the environment changed or when a new environment was entered. Introgression of genes increases the genetic and phenotypic variation of the parental species that is subject to selection (Fig. 8.8), and therefore helps to explain why morphological traits of the few studied species are so highly heritable (ch. 5). Evolutionary implications of introgressive hybridization will be considered further in the following chapters, especially chapter 11.

Summary

The barrier to interbreeding is not complete in all cases at the time that sympatry is established. Darwin's finch species hybridize rarely, sometimes as a result of young birds misimprinting on the song of another species. The circumstances under which misimprinting occurs are idiosyncratic. Interbreeding leads to the production of hybrids, which backcross to one or the other parental species according to the song type of their fathers. Introgressive hybridization demonstrates that speciation is not completed in allopatry. Survival of hybrids and backcrosses depends on ecological conditions. Under conditions favorable for birds with intermediate beak sizes, hybrids and backcrosses survive and reproduce as well as the parental species if not actually better than them. They are relatively unfit under other ecological conditions, but there is no evidence of post-zygotic genetic incompatibilities. In principle, differences between incipient species in mating signals and responses that arise in allopatry could be enhanced (reinforced) by selection against hybridization in sympatry. However, the potential for reinforcement does not seem to be strong, given the evidence of introgressive hybridization and generally high hybrid fitness when ecological conditions are favorable for survival of the hybrids. One study that looked for evidence of song divergence in sympatry found little evidence for it. Introgression of genes increases the genetic and phenotypic variation of the parental species, and could have been important in facilitating rapid evolution in the past when the environment changed.

Species and Speciation

It is really laughable to see what different ideas are prominent in
various naturalists' minds, when they speak of "species"; in some,
resemblance is everything, and descent of little weight—in some,
resemblance seems to go for nothing . . . in some, descent is the key—
in some, sterility an unfailing test, with others it is not worth a
farthing. It all comes, I believe, from trying to define the undefinable.
(Charles Darwin letter to J. D. Hooker, 1856; Darwin 1887, vol. II, p. 88).

The power of remaining for a good long period constant I look
at as the essence of a species, combined with an appreciable
amount of difference.
*(Charles Darwin letter 179 to J. D. Hooker, 1864; Darwin and Seward
1903, vol. I, p. 252).*

INTRODUCTION

POPULATIONS DIVERGE and eventually become species (Fig. 9.1). At
what point in this continuous process of change should one species be
considered as two? This chapter discusses the problem of defining and
recognizing species in the context of the adaptive radiation of Darwin's
finches. We adopt the biological species concept, provide a working definition
of species, and then apply it to two contentious issues. Strongly differentiated,
allopatric populations pose one kind of problem—they have no opportunity to
interbreed. Hybridizing populations pose the opposite problem, that of determining how much gene exchange through interbreeding is too much.

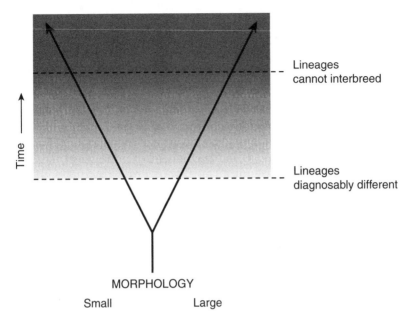

Fig. 9.1 A simple representation of speciation. Isolated populations (lineages) diverge. Opinions vary on when the lineages merit recognition as two species (see text).

From Process to Product: What Is a Species?

To a systematic biologist the first major question to be answered is not how do species evolve, but what is a species (Cracraft 2002)? Templeton (1989) has suggested the question must be answered first before the process of species formation can be investigated. The sequence seems logical and straightforward. However, the species question is a general problem without a general solution (Grant 1993, Coyne and Orr 2004). Therefore, like Darwin in writing the *Origin of Species*, we have not followed the sequence because there is no universal agreement on how to define species. Dobzhansky (1941, p. 341) explained the problem this way:

> The cause of this truly amazing situation—a failure to define species which is supposedly one of the basic biological units—is not too difficult to fathom. All of the attempts mentioned above have strived to accomplish a patently impossible task, namely to produce a definition that would make it possible to decide in any given case whether two given complexes

of forms are already separate species or are still only races of a single species. Such a task might be practicable if species were separate acts of creation, or else if species would arise from one another by a sudden, catastrophic, change, like a single mutational step.

A WORKING DEFINITION

Biologists are united in viewing species as populations that are segments of ancestor-descendent evolutionary lineages (de Queiroz 1998, p. 63), yet continue to debate how much difference between populations is enough to justify calling them separate species. Should species be simply diagnosably different (Fig. 9.1) with genetic information such as the sequence of the cytochrome oxidase I gene (Hajibabaei et al. 2006)? Species so recognized can be objectively identified and classified without regard to their interbreeding potential. Alternatively, two populations might be considered species only when they are completely incapable of exchanging genes (Dobzhansky 1935, Barton and Hewitt 1985). This alternative has the merit of being the unambiguous point of no return. Lying between these early and late extremes in the continuous process of divergence are numerous situations of occasional interbreeding without the populations losing their distinctiveness. However, it may take a long time to determine that such species persist as distinct entities despite hybridizing. Speciation is a process and not an event.

Our principle concern in this book is with the process of speciation, its causes and consequences, and not with defining its product. So far, instead of defining species, we have found it sufficient to simply invoke the biological species concept (Wright 1940, Dobzhansky 1941, Mayr 1942, Coyne and Orr 2004) with its emphasis on the evolution of reproductive isolation and not on classification (Harrison 1998, Hudson and Coyne 2002). At the end of an exceptionally long career, Ernst Mayr wrote: "I define biological species as groups of interbreeding natural populations that are reproductively (genetically) isolated from other such groups" (Mayr 2004). According to this view, what separates species are the traits that maintain that isolation, sometimes called an isolating mechanism. An isolating mechanism is defined as "a genetically determined difference between populations that restricts or prevents gene flow between them" (Futuyma 1998; see also Coyne and Orr 2004).

In the light of what we have learned from Darwin's finches, we define species similarly, but without an exclusive emphasis on genetically determined isolation. Species comprise one or more populations whose members are capable of interbreeding with little or no fitness loss. They are reproductively isolated from all other populations, either because there is no interbreeding, since members of each do not recognize each other as potential mates, or interbreeding is rare and usually results in relatively unfit (or no) offspring being produced.

This definition allows for occasional hybridization and introgression. More importantly, it recognizes that reproductive isolation can arise and become established for non-genetic reasons. This is important for Darwin's finches and many other groups of terrestrial birds because a major component of the barrier to interbreeding is not a genetic one but is learned. Finch species share the biological property of a song-learning program, indeed they may possess an identical program, but differ because they learn and produce different songs according to their different early experience. It is what they learn, not the ability to learn, that keeps them apart. For some groups of species, mainly vertebrates, cultural inheritance may be as important as genetic inheritance in maintaining biologically important traits used in reproduction, thereby perpetuating the differences between species.

In some ways song is the cultural equivalent of the W chromosome in birds (or the Y chromosome in humans): inherited from only one parent, non-recombining, and subject to change by (cultural) mutation, although far more rapidly as a result of learning than any gene on the W chromosome. To put it more generally, the environment experienced by developing organisms, and not just their genes, can influence subsequent mating patterns. This is even true of some invertebrates, such as fruit flies (Brazner and Etges 1993, Etges 1998) and spiders (Hebets 2003), which are usually thought to choose mates under strict neurological control imposed by their genes.

How Many Species of Darwin's Finches?

We now apply our working definition to the question of how many species there are. At minimum Darwin's finches constitute a group of 14 biological species. Twelve coexist with close relatives in various combinations on Galápagos islands and maintain their distinctness even if they do on rare occasions interbreed. Two more are geographically isolated from close relatives. Strongly differentiated allopatric populations pose a well-known problem for those wishing to assign species names to them, for the obvious reason that the criterion of reproductive compatibility with related populations cannot be applied without some guesswork (Zink and McKitrick 1995, Helbig et al. 2002, Newton 2003).

One of the two isolated species (*Pinaroloxias inornata*) occurs alone on Cocos island. It is genetically and morphologically so distinct that no questions have been raised about its status as a separate species. The second species, *Geospiza conirostris*, is geographically isolated on Española and Genovesa from its *G. scandens* relatives. The population on Española is sufficiently different from it genetically, morphologically, and in song (Bowman 1983, Ratcliffe and Grant 1985) to be recognized as a distinct species (Petren et al. 2005). The population of *G. conirostris* on Genovesa, however, is less distinct (Plate 20) and more problematic (Lack 1947, Petren et al. 2005). Discrimination tests with *G. conirostris* on Genovesa have only been performed with song, and they showed no discrimination between the song of *G. scandens* from Daphne Major and their own (Ratcliffe and Grant 1985). On the other hand, *G. scandens* on the island of Plaza Sur discriminated against female specimens of *G. conirostris* from Genovesa by several response measures (Ratcliffe and Grant 1983b). In the absence of similar tests on Genovesa, we have followed the conservative course of retaining the name of *G. conirostris* for the Genovesa population, recognizing inevitable subjectivity in the judgment that if it was sympatric with *G. scandens* the two would not interbreed so freely as to fuse into one species. It exemplifies stage 2 in the allopatric speciation process modeled in Fig. 3.1, and at the same time it epitomizes the allopatric species problem of taxonomy.

In addition to the 14 species recognized by Lack (1947) and by us there are three more possibilities that merit consideration: one extra warbler finch and two extra sharp-beaked ground finches.

Certhidea olivacea: *One Species or Two?*

The stem group of Darwin's finches is composed of two lineages of warbler finches, the *olivacea* and *fusca* lineages (Petren et al. 1999). They have been isolated from each other for longer (1.5–2.0 MY or more) than any other set of populations, according to microsatellite (Petren et al. 1999, Tonnis et al. 2004) and mitochondrial DNA differences (Freeland and Boag 1999a), yet in plumage (Plates 2 and 4), and especially in size, they differ in only minor ways, which shows that substantial ecological differentiation in allopatry (ch. 3) is far from inevitable. In contrast to their morphological similarity, their songs are recognizably different. Nevertheless, tape recordings from two populations of each lineage played back to each other revealed little evidence of discrimination by song (Grant and Grant 2002d). The implication is that they would interbreed frequently if sympatric. It is possible, however, that the two lineages are genetically incompatible, although we doubt this in view of apparent hybridization with small tree finches (ch. 8). While acknowledging our ignorance of post-mating compatibility, we consider the warbler finches to be well-differentiated genetic lineages of a single biological species, *C. olivacea.*

Presumably, stabilizing selection has kept populations of the two lineages in the same ecological niche, albeit in different habitats. This has retarded the evolution of a barrier to interbreeding, and stands in marked contrast to recently formed pairs of species such as the small and large tree finches. Evidently, speciation has proceeded at different rates in different parts of the evolutionary tree. The ultimate enigma is an ecological one: why has *fusca* not become established at low elevations on one of the islands that support *olivacea* populations at high elevations (Grant and Grant 2002d, Tonnis et al. 2004)? Being long resident in the archipelago, warbler finches might have become sedentary and no longer disperse from one island to another, or disperse rarely (Petren et al. 2005). Increasing sedentariness with duration of island occupancy appears to be a regular phenomenon among birds (Ricklefs and Cox 1972, Mayr and Diamond 2001). Alternatively, the habitat may be inadequate. The islands in question are Fernandina, Isabela, Santiago, and Santa Cruz. Fernandina has sparse vegetation at low elevations, naturally, and lowland habitats of the other three islands have been disturbed by humans. Unlikely though it may seem, because all the islands are large, populations of *fusca* on one or more of these islands may have become extinct.

Geospiza difficilis: *One Species or Three?*

The striking feature of *G. difficilis*, unlike *C. olivacea*, is the large morphological variation among populations. The highland populations on Pinta, Santiago, and Fernandina are all similar morphologically, genetically, and in song, and differ in all three respects from the three lowland populations. Birds on

Genovesa are particularly small (Plate 13). Those on Wolf (Plate 14) and Darwin are large and have *scandens*-like elongated beaks (Grant et al. 2000). There are thus three groups. We have nevertheless adhered to the traditional treatment of all six populations as members of the same species because of their interbreeding potential.

In playback experiments *G. difficilis* on Genovesa responded to songs of *G. difficilis* on Pinta and Darwin, which suggests that sympatry would not be established on Genovesa if *G. difficilis* immigrated from Pinta (highland group) (Ratcliffe and Grant 1985) or Darwin (lowland group) (Grant and Grant 2002b). Instead, interbreeding would result in the immigrants becoming absorbed into the resident *G. difficilis* population. In contrast, *G. difficilis* from Wolf have acquired sufficient vocal differences to be largely ignored. Nevertheless, they are morphologically almost identical to their Darwin relatives, and for this reason they are likely to interbreed on Darwin (or on Wolf). Thus it appears that each of the six populations is potentially connected reproductively with at least one other. Moreover, it is doubtful if the ones that are apparently not connected, such as those on Genovesa and Wolf, would be able to coexist. The feeding niche of *G. difficilis* on Wolf (and Darwin) is occupied by two species on Genovesa, the smaller *G. difficilis* and the larger *G. conirostris*. In both these cases the few immigrants would be at a competitive disadvantage to the residents (Grant et al. 2000).

From Product Back to Process

Counting number of species tends to obscure the fact that the boundaries of young species are fuzzy and neither sharply defined by isolating mechanisms not fixed. Here we return to the dynamic nature of species.

The ecological influence on hybrid fitness (ch. 8) has an interesting implication: with a change of the environment the divergent process of speciation can be arrested or even put into reverse. This is not a new idea: "introgressive hybridization may lead to obliteration of the differences between the incipient species and their fusion into a single variable one, thus undoing the result of the previous divergent development" (Dobzhansky 1941, p. 350). A possible example involving two species of kingfishers, *Ceyx erithacus* and *C. rufidorsus*, has been described on four islands in Indonesia (Sims 1959), and another involving two species of bulbuls (*Pycnonotus sinensis* and *P. taiwanensis*) has

114

been described on Taiwan (Severinghaus and Kuo 1994). Fusion is usually associated with anthropogenic disturbance of the environment (Cade 1983, Seehausen et al. 1997, Taylor et al. 2006).

On Daphne Major *G. fortis* and *G. scandens* have converged morphologically and genetically as a result of introgressive hybridization over a period of 20 years (Fig. 8.6). The species are thus on a reversed course of speciation. The direction could easily be reversed again, from convergence to divergence, if there is a change in climate and food composition (ch. 8). The populations are still largely kept apart by song and by their mating within song groups; therefore they function as two biological species. Their members can still be identified by us and by the finches themselves by the song they or their mates sing, as well as by beak morphology in almost all instances. They do not hold mutually exclusive territories as they would if they perceived each other as members of the same species; instead, their territories overlap. At present we continue to refer to them as two species, recognizing that they could lose their specific distinctness on Daphne Major (although not on other islands) if many individuals in each population recognize both song types, associated with increasingly similar morphology, as their own. If that happens the point at which they should no longer be considered two species, but only one, is arbitrary.

Although closely related, having shared a common ancestor in the last million years, *G. fortis* and *G. scandens* are not sister species (Plate 1 and Fig. 2.1). This shows that phylogenetic relatedness is less important in determining whether species will interbreed or not than the magnitude of the difference in song and morphology: comparison of the two *Certhidea* lineages indicates the same. As discussed in the previous chapter, *G. fortis* has hybridized on Daphne Major island with *G. scandens* but not with *G. magnirostris*, a species that is more closely related to *G. fortis* but morphologically more different.

Such fluidity and fuzziness of species boundaries is to be expected during the process of speciation (Price 2007) and especially in young adaptive radiations. Darwin's finches are valuable to evolutionary biologists precisely because they so well exemplify the graded nature of many evolutionary transitions. They are challenging to systematists for the same reason (Zink 2002). Yet, despite the fuzzy boundaries in a few cases, as many as 10 species occur on the same island (e.g., Santa Cruz), ecologically differentiated, morphologically recognizable, and reproductively isolated by song and morphology (Grant 1999).

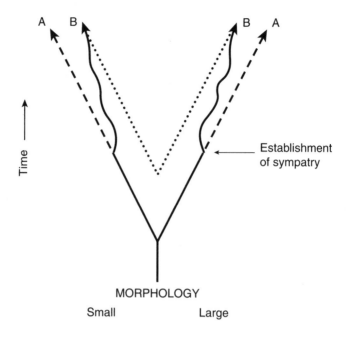

Fig. 9.2 Oscillating divergence and convergence during speciation as a result of episodic introgressive hybridization (B) contrasted with unchanging divergence as in Fig. 9.1 (A). The dotted lines show that extrapolating back to the point of origin can give a false impression of recent ancestry of hybridizing species.

FISSION AND FUSION

The process of speciation is often viewed as a steady increase in differences between populations leading eventually to the complete cessation of interbreeding. However, the introgressive hybridization and convergence witnessed on Daphne Major imply a dynamic tension between ecologically differentiated species that are derived from a common ancestor and differ reproductively only in how they choose mates. An oscillation between tendencies to merge or diverge (Fig. 9.2) characterizes the process of speciation better than uniform divergence (Fig. 9.1) or acceleration in divergence just after the time that sympatry is established. Ecological factors determine whether the oscillations are weak or strong, frequent or rare (ch. 8).

We may be witnessing merge-and-diverge dynamics in medium ground finches (*G. fortis*) on the south side of Santa Cruz island. When studied in the

1960s, beak sizes were bimodally distributed (Ford et al. 1973). Bimodality had not recently arisen; there are signs of it in the sample of specimens collected for museums more than 100 years ago (Fig. 9.3). The main mode was at a typical size, and the minor one occurred at especially large size. In the 1970s we found bimodality had disappeared at that locality, and it has not reappeared since (Hendry et al. 2006). Merging of the two modes has been attributed to natural selection caused by human alteration of the habitat and food supply in some unknown way, because 11 km away a population in less disturbed habitat has retained its bimodality (Hendry et al. 2006).

Bimodality could have arisen in three ways, separately or in combination (Grant 1999). The first is through sympatric divergence under disruptive selection (Ford et al. 1973). This would imply environmental change towards a relative scarcity of intermediate food resources, converting the usual regime of stabilizing or oscillating directional selection into a disruptive one. The second is through recent immigration of large *G. fortis* from another island. The prime candidate is San Cristóbal (Fig. 1.1), because *G. fortis* are larger on this island than anywhere else (Grant et al. 1985), and the largest individuals (beak width 12–13 mm) are as large as the largest on Santa Cruz (Fig. 9.3). This possibility conforms to the allopatric model in Fig. 3.1. The third is through hybridization with the relatively rare large ground finch, *G. magnirostris*. A point in favor of this explanation is that the songs of particularly large male *G. fortis* individuals on Santa Cruz sound similar to the songs of *G. magnirostris* (to our ears), and sonagrams published by Bowman (1983) and Huber and Podos (2006), as well as our own unpublished ones, show the motifs are similar. Song similarity could be due solely to heterospecific song copying, but is more likely to be the result of hybridization and backcrossing (ch. 8). It is also possible, in light of the recent colonization of Daphne by *G. magnirostris* from Rábida or Santiago (Grant et al. 2001), that small *G. magnirostris* individuals from Rábida (Fig. 1.3) immigrated to Santa Cruz and hybridized with *G. fortis*.

Regardless of its origin, the exceptional variation in this population (Figs. 1.3 and 9.3) could be maintained by both persistent introgressive hybridization and disruptive selection. The population has the potential to split into two reproductively isolated populations: speciation in sympatry (Ford et al. 1973). For this to occur assortative mating on the basis of song would have to arise, accompanied in all probability by territorial tolerance and mutual indifference. There would also have to be enough ecological niche space for the new species, a *magnifortis*, between the niches of typical *G. fortis* and *G. magnirostris*. We

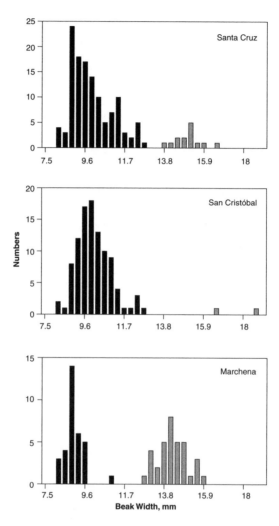

Fig. 9.3 A conspicuously skewed distribution of beak sizes of *G. fortis* (black) on Santa Cruz (upper). One explanation for this is immigration (and breeding) of large birds of this species from San Cristóbal (middle). On other islands, such as Marchena (lower), the largest *G. fortis* individuals are smaller in average size. Another explanation is interbreeding of *G. fortis* on Santa Cruz with the much rarer *G.magnirostris* (grey), and backcrossing of the hybrids to *G. fortis*. The same may have happened on San Cristóbal; the two *G. magnirostris* individuals on San Cristóbal were collected by Darwin and colleagues in 1835, and are the remnants of the largest form of this species. Soon after 1835 the population, and a sister population of large birds on Floreana, became extinct. Interbreeding is less likely on islands where numbers of the two species are

doubt the space exists. The incipient species would be hemmed in, as it were, between morphological (Fig. 1.3) and ecological neighbors (ch. 6), unless *G. magnirostris* became extinct. It would have a potentially richer future if some individuals dispersed to a small island and founded a new population, as *G. magnirostris* did on Daphne Major (ch. 4). This intriguing situation is currently being studied (S. K. Huber and J. Podos pers. comm.).

SUMMARY

This chapter addresses the question of when diverging populations should be considered as two species. There is no universal agreement on how species should be defined. Two populations might be considered species at an early stage of divergence when they are diagnosably different, or at the end of speciation when they are completely incapable of exchanging genes. Lying between these extremes are numerous situations of occasional interbreeding without the populations losing their distinctiveness. Taking this into account, we adopt the biological species concept in framing a two-sentence definition. Species comprise one or more populations whose members are capable of interbreeding with little or no fitness loss. They are reproductively isolated from all other populations, either because there is no interbreeding, or because any interbreeding is rare and usually results in relatively unfit (or no) offspring being produced.

This definition allows for occasional hybridization and introgression, and it recognizes that reproductive isolation can arise and become established for non-genetic reasons. The process of speciation is often viewed as a steady increase in differences between populations leading eventually to the complete cessation of interbreeding. However, the introgressive hybridization and convergence of *G. fortis* and *G. scandens* witnessed on Daphne Major island imply a dynamic tension between ecologically differentiated species that are derived from a common ancestor and that differ reproductively only in how they choose mates. An oscillation between divergence and convergence as a result of changing ecological conditions might characterize the process of speciation better than uniform divergence or acceleration in divergence just after the time that sympatry is established.

---

more equal, as on Marchena (note one intermediate individual, however; possibly a hybrid if not an immigrant from another island).

Reconstructing the Radiation of Darwin's Finches

This is the process grandly exemplified in the thinning out and
increasing divergence and specialization of the various lines of
descent in an adaptive radiation.

(Simpson 1949, p. 208)

INTRODUCTION

ADAPTIVE RADIATION is as much an ecological as an evolutionary
phenomenon (Schluter 2000), as implied by the adjective "adaptive."
It results in an accumulation of ecologically diverse species through
repeated speciation. In chapter 3 we referred to a graphical representation of
speciation (Fig. 3.1) as an abstraction designed to capture the essence of the
process from a mass of particulars. In this chapter we pay attention to the
particulars that have been left out so far. They vary from one speciation cycle
to another, as does the environment, which partly accounts for the diversity
and helps us to understand why, for example, tree finches were ultimately
derived from warbler finches and not vice versa.

As described in chapter 2, the Galápagos environment has undergone
fluctuations in temperature and sea level, but with long-term trends of cooling,
aridification, and increase in number of islands. The radiation of finches
unfolded with an increase in number and diversity of species in a changing
environment: an increase in number of islands increased the opportunities for
speciation and thereby the *number* of species (Figs. 2.2 and 2.4), and a change
in climate and altered vegetation increased the opportunities for *new types* of
species to evolve. The net result, the product of the evolutionary process, is a
heterogeneous array of species in a heterogeneous archipelago; the larger and
higher the island, the more habitats it supports and the more numerous and
diverse are the finch species (Abbott et al. 1977).

Since the exercise of interpreting the past is retrospective with limited scope for verification, several aspects of the radiation raise more questions than can be answered at present. Moreover, a study of adaptive radiation without fossils is a study of the survivors, not the whole of the radiation, unless there has been no extinction, which seems very improbable in the light of the fossil record in general (Valentine 1985, 2004). Episodic extinction is almost inevitable in a seismically and volcanically active environment like the Galápagos. Missing data on extinctions are the largest limitation on what can be achieved in attempts to interpret evolutionary history. Extinctions have been aptly described by Williams (1969) as "invisible history."

Discussion of some parts of the radiation is therefore unavoidably speculative. We view the past through the lens of the present, and the lens is clouded and distorted. Using what is known from paleoclimatology, geology, evolution, ecology, and biogeography, we do our best to recreate the past in order to understand the present. We adopt a uniformitarian principle from geology in assuming that processes occurring in the past were fundamentally the same as those we observe today—natural selection, adaptation, competition, and introgressive hybridization. Environmental circumstances were different, however.

THE SHAPE OF THE RADIATION

Species that evolved early in the radiation differ in three ways from those that evolved relatively late: they display (a) almost the complete generic, morphological (Fig. 10.1) and ecological diversity of the whole group, yet (b) no species diversity within genera, and therefore (c) no possibilities for sympatry within genera. The early set of species comprises a warbler finch species (*Certhidea*), the Cocos finch (*Pinaroloxias*), one ground finch (*G. difficilis*), and one vegetarian finch (*Platyspiza*). Five tree finch species and an additional five ground finch species constitute the late set, and they occur in various sympatric combinations.

The sharp contrast between the early and late sets gives the radiation a structure. The structure is not expected from a simple model of species accumulation with time. In such a model the earliest species would have the greatest opportunity to produce new species similar to themselves. For example, there are five species of tree finches, hence there has been enough time for at least five (older) vegetarian finch species to evolve, differing in beak size or

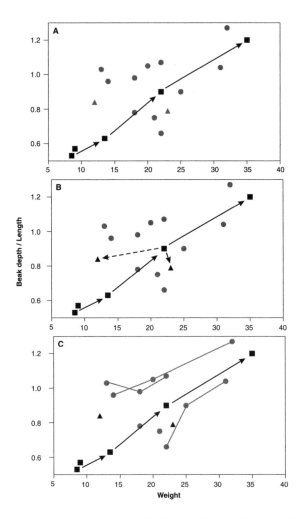

Fig. 10.1 Accumulation of morphological diversity (disparity) among Darwin's finch species on the Galápagos in three stages, A, B, and C. Upper: the early set of species span the full range of body sizes and beak shapes. The species are connected by a line on the basis of their microsatellite DNA relationships and branching points in Fig. 2.1; arrows give an evolutionary interpretation of the time course. Middle: the central point represents the highland populations of *G. difficilis*, and is connected with broken arrows to later evolving conspecific populations on the two northernmost islands and on Genovesa. Lower: triplets of related species in the late set of species are connected by lines to show that their variation is roughly parallel to the early set. For a given body size they have relatively pointed or robust beaks. Beak proportions have been calculated from mean measurements of males from Lack (1947). Weights are from Grant et al. (1985).

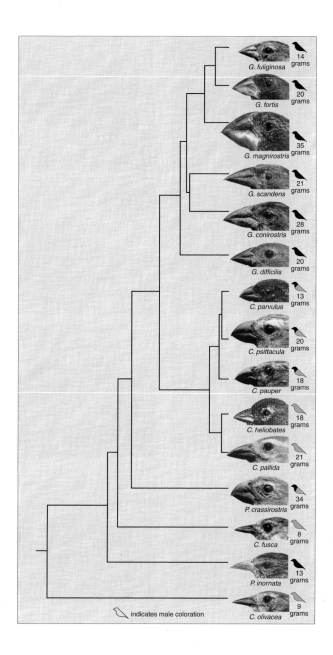

1. Relationships among Darwin's finch species constructed by using the differences among them in microsatellite DNA. Suport for the nodes has been omitted. From Petren et al. (1999) and Grant and Grant (2002a). Most populations of *G. difficilis* originated earlier according to the revised version in Figure 2.1.

(all photos are by the authors except where otherwise noted)

2a. Darwin's finch species

a. Warbler finch, Certhidea olivacea (olivacea lineage)
b. Warbler finch, *Certhidea olivacea* (*fusca* lineage)
c. Woodpecker finch, *Camarhynchus pallidus*
d. Mangrove finch, *Camarhynchus heliobates*
e. Vegetarian finch, *Platyspiza crassirostris*
f. Small tree finch, *Camarhynchus parvulus*
g. Medium tree finch, *Camarahynchus pauper*
h. Large tree finch, *Camarhynchus psittacula*

2b. Darwin's finch species

- **i.** Sharp-beaked ground finch, Geospiza difficilis (Santiago)
- **j.** Sharp-beaked ground finch, *Geospiza difficilis* (Genovesa)
- **k.** Cactus finch, *Geospiza scandens*
- **l.** Large cactus finch, *Geospiza conirostris*
- **m.** Large ground finch, *Geospiza magnirostris*
- **n.** Medium ground finch, *Geospiza fortis*
- **o.** Small ground finch, *Geospiza fuliginosa*
- **p.** Cocos finch, *Pinaroloxias inornata*

3. Darwin's finch relatives

Upper: *Melanospiza richardsoni* (Saint Lucia island, Caribbean; J. Faaborg); **middle:** *Tiaris olivacea* (Panama; M. Wikelski); **lower:** *Certhidea olivacea* (Santiago island, Galápagos).

4. Warbler finches from four islands

Upper left: Santa Cruz (*olivacea* lineage) (R. Å. Norberg); **upper right:** Española (*fusca* lineage) (R. I. Bowman); **lower left:** Genovesa (*fusca* lineage) (O. Jennersten); **lower right:** San Cristóbal (*fusca* lineage) (K. Petren).

5. Upper: Daphne Major (D. Parer and E. Parer-Cook); **lower:** volcanic activity on Sierra Negra, Isabela island, Galápagos, 1979 (M. F. Kinnaird).

6. Cocos island (**above**, N. Grant) and its finch (**below**).

7. FOUR GALÁPAGOS HABITATS AT DIFFERENT ELEVATIONS

Upper left: Lowland (San Cristóbal); **upper right:** Transition (Pinta); **lower left:** *Zanthoxylum* (Pinta); **lower right:** *Scalesia* (Santa Cruz).

8. Capturing, measuring, and banding birds on Daphne Major

Upper left: Cactus finch in a mist net; **upper right:** weighing and taking a blood sample (K. T. Grant); **lower left:** weighing a bird; **lower right:** banded large ground finch (three color bands and no metal band because large birds can crush them).

9. MEASUREMENTS OF A LARGE CACTUS FINCH ON GENOVESA

Upper left: Beak length; **upper right:** Beak depth; **lower left:** Beak width; **lower right:** Tarsus length.

10. THE FOUR SPECIES OF DARWIN'S FINCHES ON DAPHNE MAJOR

Upper left: Small ground finch; **upper right:** Medium ground finch; **lower left:** Large ground finch; **lower right:** Cactus finch.

11. LARGE TREE FINCHES, DIFFERING IN BEAK SIZE AND SHAPE ON TWO ISLANDS

Upper: Pinta; **lower:** Isabela (Alcedo).

12. Habitats of lowland (left) and upland (right)
populations of the sharp-beaked ground finch

Upper left: Darwin (M. Wikelski); **middle left:** Wolf; **lower left:** Genovesa; **upper right:** Fernandina (with mist net); **middle right:** Santiago (vegetation decimated by goats that have now been removed); **lower right:** Pinta.

13. SHARP-BEAKED GROUND FINCHES

Upper: Genovesa, a low island (R. L. Curry); **lower:** Pinta, a high island (D. Nakashima).

14. Unusual feeding habits
of the sharp-beaked ground finch on Wolf

Upper: Inspecting the egg of a booby (*Sula* sp.); **middle:** Egg-rolling; **lower:** Blood-drinking from the base of wing feathers. All photos by D. Parer and E. Parer-Cook.

15. PLANTS THAT PRODUCE SMALL SEEDS ON DAPHNE MAJOR

Upper left: *Chloris virgata*; **upper right:** *Eragrostis cilianensis*; **lower left:** *Heliotropium angiospermum*; **lower right:** *Portulaca howelli*.

16. Feeding of medium ground finches on Daphne Major

Upper left: Fruits of *Chamaesyce amplexicaulis*; **upper right:** Capsules of *Sida salviifolia*; **lower left:** Seeds and nectar of *Sesuvium edmonstonei;* **lower right:** Spiderlings.

17. EFFECTS OF A DROUGHT ON DAPHNE MAJOR VEGETATION

Upper: *Chamaesyce amplexicaulis*; **lower:** Prolonged drought results in the death and disappearance of plants.

18. *Tribulus cistoides*

Upper: Flowering plant; **lower:** Fruits.

19. EFFECTS OF EL NIÑO ON DAPHNE MAJOR VEGETATION

Upper left: Dry season (*Bursera graveolens* trees are leafless); **upper right:** Wet season (trees are in leaf); **middle left:** Extensive growth of annual plants in early stage of El Niño development; **middle right:** Late stage of El Niño; **lower left:** Cactus bush almost entirely smothered by *Merremia aegyptica* vines; **lower right:** Aftermath of El Niño (dead vines drape cactus bushes and trees).

20. THE LARGE CACTUS FINCH IN THE PRESENCE AND ABSENCE
OF THE LARGE GROUND FINCH

Upper left: Large cactus finch, Genovesa (cactus finch absent); **upper right:** Large cactus finch, Española (large ground finch and cactus finch absent), female left, male right; **lower left:** Cactus finch; **lower right:** Large ground finch.

21. FEEDING OF LARGE GROUND FINCHES ON DAPHNE MAJOR

Upper left: Fruit of *Croton scouleri;* **upper right:** Seed capsule of *Portulaca howelli*; **lower left:** Fruit of *Tribulus cistoides*; **lower right:** Mericarp of *Tribulus cistoides*.

22. Cactus fruits exploited by cactus finches

Upper: *Opuntia helleri* on Genovesa (seeds have been eaten by the large cactus finch and inedible protective coverings have been discarded on the pad); **middle:** *Opuntia echios* on Daphne Major exploited by cactus finches (the seeds are smaller and so are the beaks of the finches on this island); **lower:** Opened and discarded seeds of *O. echios* on Daphne Major.

23. CACTUS FEEDING BY CACTUS FINCHES ON DAPHNE MAJOR

Upper: Opening a bud; **middle:** Feeding on pollen and nectar; **lower:** Feeding at the base of spine clusters.

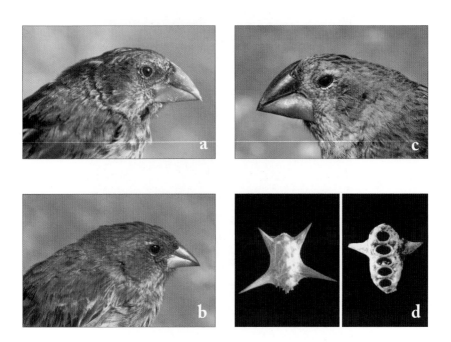

24. CHARACTER DISPLACEMENT ON DAPHNE MAJOR

Large members of the medium ground finch population (a) died at a higher rate than small members (b) owing to depletion by large ground finches (c) of the large and hard fruits of *Tribulus cistoides* (d).

25. Dead finches on Daphne Major at the end of a drought

Upper: Large ground finch (1990); **middle:** Cactus finch, Medium ground finch and large ground finch (2004); **lower:** About 50 individuals of all species (2004).

26. EXPERIMENTAL SETUP TO TEST DISCRIMINATION OF FINCHES
BY APPEARANCE AND BY SONG

Upper: A stuffed specimen is placed at each end of a rod mounted on a tripod and placed in a finch's territory; **middle:** Response of a male sharp-beaked ground finch on Genovesa to a female in copulation soliciting posture; **lower:** Loudspeaker for testing responses of finches on Genovesa to playback of tape-recorded song.

27. A probable hybrid on Genovesa **(middle)** produced by a large cactus finch **(upper)** breeding with a large ground finch **(lower).**

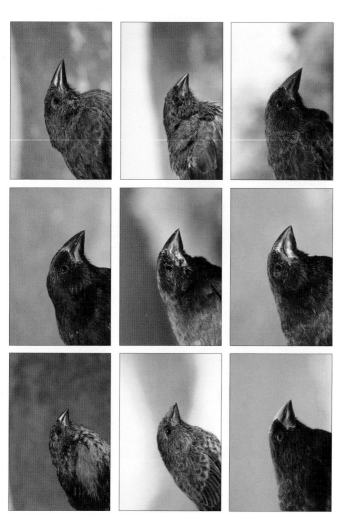

28. Hybrids and backcrosses on Daphne Major

Top row: Representatives of the three hybridizing species, small groundfinch (**left**), medium ground finch (**center**), and cactus finch (**right**); **middle row:** F₁ hybrids, small x medium ground finch (**left**) and cactus x medium ground finch (**center and right**); **bottom row:** First-generation backcrosses, small x medium groundfinch backcrossed to medium ground finch (**left**) and cactus x medium groundfinch backcrossed to medium ground finch (**center**) and to cactus finch (**right**).

29. Unusual feeding habits

Upper: Use of a tool to extract insect larvae from wood by a woodpecker finch; **middle:** Tick-eating by a small ground finch from a marine iguana; **lower:** Same from a land iguana. All photos by D. Parer and E. Parer-Cook.

30. FOUR ALLOPATRIC SPECIES OF GALÁPAGOS MOCKINGBIRDS

Upper left: Española; **upper right:** Champion; **lower left:** Genovesa; **lower right:** San Cristóbal (R. L. Curry).

31. Upper: The vangids of Madagascar (S. Yamagishi and K. Kanao). Species names are given in Yamagishi and Honda (2005); **lower:** The honeycreeper finches of Hawaii (H. D. Pratt). Painting © H. Douglas Pratt.

shape and diets and coexisting in various combinations. And yet there are no sympatric vegetarian finches, or warbler finches, or sharp-beaked ground finches, or Cocos finches. How can we account for the differences between the early and late set of species?

The singular Cocos finch is a special case and is easily explained. Isolated on a single island, it has no opportunity to diversify in separate locations. The species has been present long enough to have given rise to other species (Fig. 2.1), and its environment is varied enough to support a variety of feeding types in the population (Werner and Sherry 1987), and yet it has remained a single species under conditions suitable for sympatric, but not allopatric, speciation. These facts have been used as an argument against sympatric speciation of Darwin's finches (Lack 1947, Grant 1999, Coyne and Price 2000).

SPECIATION AND EXTINCTION

Accumulation of species is the net result of two opposing processes, speciation and extinction. If speciation and extinction can be likened to birth and death processes within populations, and their rates are assumed to be constant but not necessarily equal in a given time interval, one can use the temporal pattern of species origins derived from molecular data to see if overall speciation has proceeded at a constant or differing rate (Nee 2006, Rabosky 2006, Weir 2006, Price 2007). The Darwin's finch radiation, analyzed in this way, has been temporally heterogeneous: assuming extinction rate to have been constant, speciation was slower in the first half than in the second (Schluter 2000). Figure 10.2 illustrates the accumulation of species. After the warbler finches split into two lineages there was an apparently long time before the next two Galápagos species, the sharp-beaked ground finch and vegetarian finch, were formed.

Possible causes of low speciation rate early in the radiation are not difficult to identify. The archipelago was less heterogeneous then than now: fewer islands, less habitat diversity, fewer food types. It is more difficult to interpret the strong morphological and ecological differences among the early set in terms of contemporary evolution. In contrast, the late evolving set are easy to understand because species transformations among them were small. We discuss effects of low speciation and high extinction on the early set, first separately and then together.

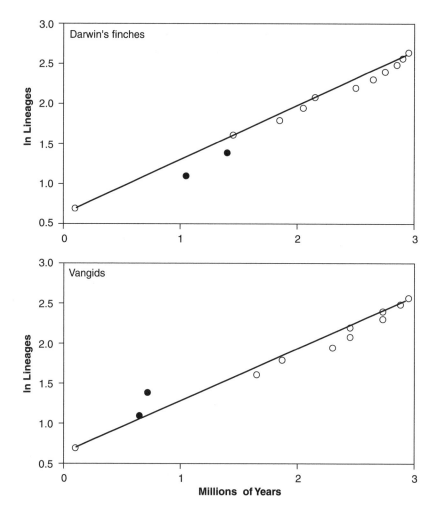

Fig. 10.2 Accumulation of species through time. After the first division of Darwin's finches into two lineages of warbler finches (bottom left in upper diagram), an unexpectedly long time appears to have elapsed before the origin of the next two species on Galápagos; points for these two (solid symbols) fall below the line drawn to illustrate a regular exponential increase in number of species. The radiation of vangids on Madagascar (lower diagram), discussed in the next chapter, displays a contrasting pattern of accumulation of species that is more rapid early than late. Based on Fig. 2.1 and RNA sequence data in Fig. 8–6 of Yamagishi and Honda (2005).

Speciation

Warbler finches, vegetarian finches, and highland sharp-beaked ground finches occupy very different niches, far apart in a hypothetical landscape of adaptive peaks and valleys. At one time under warm and wet conditions (ch. 2) these may have been the only species of finches on the Galápagos. Morphological diversification could have proceeded very rapidly and extensively as is often claimed for other radiations (e.g., Foote 1997, Harmon et al. 2003, Valentine 2004, Gavrilets and Vose 2005, Ruta et al. 2006, Seehausen 2006; see also the vangids in Fig. 10.2), but not accompanied by much speciation. Or diversification was slow and gradual, episodic rather than continuous (chs. 5 and 6), through long-term adaptive change within each lineage after reproductive and ecological isolation had evolved—so-called phyletic evolution beyond speciation (Simpson 1949). Thus, for example, vegetarian finches might have acquired their present, distinctive features either a long time ago, rapidly, or more recently, gradually.

The problem posed by both alternatives but especially by the first is to explain the large morphological gaps between species: why did evolutionary diversification not stop after a minimally sufficient difference for coexistence had arisen? After all, pronounced diversification is not inevitable, as shown by the minimal difference in morphology between the two *Certhidea* lineages. Lacking a detailed knowledge of paleocommunities of plants and arthropods, we have no answer to the question. The second alternative of gradual change seems more plausible than the first when taking into account how ecological communities generally develop by speciation and immigration (Ricklefs and Schluter 1993).

Immigration of new plants and arthropods would have aided both speciation and directional adaptive change after speciation by periodically producing new ecological opportunities for consumers. Perkins (1903) offered a similar explanation for the buildup of the honeycreeper finch fauna in the Hawaiian archipelago in terms of evolutionary responses to opportunities provided by the radiation of lobelias and an expansion of the insect fauna. On the Galápagos, climatic change permitted the establishment of arid-adapted plants, and their inferred influx provides a plausible explanation for the high speciation rate of tree finches and ground finches. The highland populations of the sharp-beaked ground finch played a pivotal role in this expansion (Fig. 10.1, middle). Originating relatively early in the history of Darwin's finches in *Zanthoxylum* forests, and somewhat generalized in morphology, they gave rise to new, arid-adapted populations (Fig. 2.1) as well as to a mini-radiation of ground finch species, possibly also to the tree finches. Morphologically these new species diverged above and below the initial evolutionary trajectory towards blunt beaks and large size.

These ideas on the effects of new immigrants need to be tested with molecular phylogenies of key plants such as *Zanthoxylum* and *Opuntia*. There are no molecular phylogenies of Galápagos plants. All that is known at present is that *Opuntia* arrived fairly recently: Ecuadorean mainland and Galápagos *Opuntia* scarcely differ in allozymes (Browne et al. 2003).

Extinction

Extinction of species with intermediate morphologies provides an explanation for wide morphological gaps between species and for the slow rate of species accumulation in the early part of the Darwin's finch radiation. High rates of extinction among the early-formed species could give a spurious impression of slow speciation, and invalidate the approach used above. The paleontologist G. G. Simpson interpreted repeated patterns in the fossil record as a result of a "thinning out" of the species first formed in a radiation (see quotation at head of chapter). The modern view from the world of fossils is that extinction is both diversity-dependent, and environment-dependent. It tends to increase as the diversity of species increases, and is especially pronounced when the environment changes (Valentine 2004).

Extinction of some members of the early set could have been caused by habitat change associated with increasing aridity. Competition, both among the early species and with the tree and ground finch species that proliferated in the late phase, could have contributed to extinction. The later species are likely to have been more versatile ecologically, because their environments were more unpredictable, and more adaptable to a changing constellation of foods. For example, anthropogenic removal of the *Zanthoxylum* forests on the islands of Santa Cruz and Floreana resulted in the extinction of sharp-beaked ground finch populations that were largely restricted to forests dominated by this species (Grant 1999). Other ground finch species and tree finches now exploit the transformed and degraded habitat on these islands.

Extinction in recent times sheds a little light on extinctions in the past. A change in habitat is likely to have been the cause of the only known extinction of a Darwin's finch population in the absence of human influence. The population of mangrove finches restricted to small patches of mangroves on the east coast of Fernandina became extinct in the twentieth century (Grant and Grant 1997d). The apparent cause was seismic activity, land elevation, drying of the soil, and death of the favored old individuals of white mangroves and black mangroves. There is no reason to believe that competition with other finch species was involved. For example, there are no tree finches in the mangroves on this island. Thus extinction of populations has occurred in modern times when habitat has been lost either naturally or anthropogenically (Grant et al. 2005b).

Putting these ideas together, we suggest that changing habitat conditions and competitive exclusion jointly eliminated the species in the early set most similar ecologically to those in the later set. Some species became extinct through loss of the resources they were adapted to exploit during the transformation of Galápagos vegetation from rainforest to current, heterogeneous, and partly arid-adapted, forest. The early species surviving to the present are ecologically and morphologically the most different from the late species and from each other: the smallest species of a possible set of warbler finches and the largest of a possible set of vegetarian finches (Fig. 10.1). As a consequence there is only one species per genus, hence no sympatric combinations of congeneric species.

Implications for Phylogeny

The likelihood of extinction casts a different light on the phylogeny. The representation in Figure 2.1 is the simplest estimation of history, given the genetic data of existing species. In other words it is visible history, and it assumes no distorting influences from the invisible history of missing species (extinction). However, as we have suggested, extinction of some of the early-formed species is probable, and the *Tiaris*-like ancestral species (ch. 2) itself may have been one of the casualties.

Taking extinction into consideration we suggest the following sequence. The first new species to evolve on one of the five Galápagos islands was the warbler finch. A second species originated from one of the ancestral populations, and not from the warbler finch. It was either the species that later became the Cocos finch—a proto-Cocos finch—the sharp-beaked ground finch, or a species that gave rise to both. Cocos Island was then colonized, and some time later the source population on Galápagos became extinct. Its extinction on Galápagos is sufficient to explain the apparently long lag in speciation after the warbler finch was formed (Fig. 10.2).

The suggested relationships make sense of the plumage features of the Cocos finch, which are black in the male and brown in the female, as in *Tiaris*, other continental and Caribbean relatives, and ground finches, but unlike the green warbler finches. Furthermore, the Cocos finch and some populations of sharp-beaked ground finches share a unique rusty color in wing and under-tail coverts (Lack 1947), and have similar songs (Grant et al. 2000). In terms of microsatellite DNA, the genetically most similar populations to the Cocos finch are sharp-beaked ground finches on Pinta and Santiago.

This revised view of early phylogeny conflicts with the apparent derivation of all species from the *fusca* lineage of warbler finches. The conflict is resolved by relaxing the assumption of genetic independence after two species separated from one, an assumption that is clearly violated by hybridization of modern species (Figs. 8.6 and 8.7) and encouraged by depiction of their relationships in sharply dichotomous branching patterns (Fig. 2.1). Introgressive hybridization, a second element of invisible history, can distort the reconstruction of history if gene exchange is asymmetrical, and if some pairs of species exchange genes more than others. History is most prone to distortion when estimated from mitochondrial DNA, because it is only a single molecule, maternally inherited. We suggest introgression of alleles occurred from the *fusca* lineage of warbler finches into the proto-Cocos finch and the sharp-beaked ground finch early in their history, causing some degree of genomic blending and reticulate evolution. Phylogenetically, one lineage captured another through introgression (Fig. 10.3).

The apparently anomalous position of the Genovesa population of sharp-beaked ground finches in the phylogeny may have a similar explanation: it may be the result of a combination of a high rate of extinction of earlier species along one branch, a high rate of speciation recently, and gene exchange among the products (Figs. 2.1 and 5.4).

Adaptive Landscape

Our thesis is that changes in resources, and competition for them, opened up opportunities for speciation but also caused some extinctions. Jointly they helped to give the radiation its shape. The conceptual framework of an adaptive

PHYLOGENY A : some history is invisible

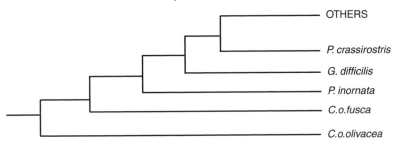

PHYLOGENY B: history revealed

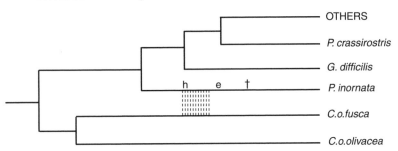

Fig. 10.3 Phylogeny with hybridization and extinction. Phylogeny A depicts the conventional representation of relationships among early-formed species of Darwin's finches in a branching diagram based on genetic similarities (Fig. 2.1). Actual branching sequences may have been different. Phylogeny B illustrates one possibility. The Cocos finch (*P. inornata*) may have hybridized (h) with one group of warbler finches, as indicated by broken lines, before some individuals emigrated (e) and colonized Cocos island; extinction (†) of the species on Galápagos occurred later. As a result *P. inornata* appears in phylogeny A to have evolved from *C. olivacea*.

topography or landscape (Fig. 10.4) exemplifies the main ideas and allows further exploration of them.

Sewall Wright (1932) initially constructed the landscape in terms of gene combinations, then Simpson (1944) modified it to deal with phenotypes. The basic idea is that the N-S and E-W axes of a landscape represent variation in two morphological characters suitable for exploiting food or other resources. The third, vertical, axis is fitness. There are fitness peaks in the landscape

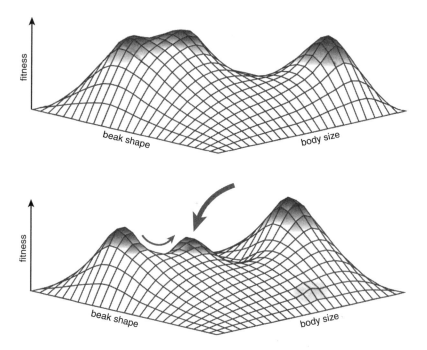

Fig. 10.4 An adaptive landscape, in which various combinations of body size and beak shape confer different fitnesses on individuals in a single island environment. A landscape with two occupied fitness peaks (upper) is converted to one with three peaks (lower) by a change in food supply. The new peak in the middle can become occupied in three ways: (a) the original population on the left-hand peak splits into two, and disruptive selection maintains them as two separated sub-populations (peak split), (b) colonization occurs from the left-hand peak (shown by a thin arrow) by a combination of genetic drift and selection (peak shift), and (c) colonization occurs by immigration from another island, as in the allopatric model of Fig. 3.1 and shown by a thick arrow (peak colonization). For this process to result in a distinct species on the middle peak, mating must be restricted to the occupants of each peak, either from the outset (c) or through evolution (a and b). If the landscape changes in reverse of the sequence shown, i.e., from lower to upper, then introgressive hybridization may be enhanced, as happened on Daphne (ch. 8), or a bimodal population becomes unimodal (ch. 9; but see ch. 3). Based on Grant and Grant (2002a).

owing to non-uniform distributions of resources and optimal combinations of morphological character values for exploiting them. Ascent of a population to a peak in fitness in the adaptive landscape is caused by natural selection; therefore, valleys between two peaks represent a barrier. Wright invoked random genetic drift to account for the crossing of the valley. Alternatively, according to the speciation model in Fig. 3.1 a new peak is colonized by immigrants from an island with a somewhat different adaptive landscape.

The adaptive landscape has been made operational by using seed resources on several islands to construct a general relationship between maximum density profiles (fitness) and beak sizes for all of the granivorous species of Darwin's ground finches combined (Fig. 10.5). Then, mean beak sizes of populations on each of 15 islands were predicted from peaks in the density profiles that were generated from resource distributions on each of those islands (Schluter and Grant 1984a, Schluter et al. 1985). There were two main results from this exercise. First, with few exceptions, observed average beak sizes were generally close to what they were predicted to be. Second, no more than one species was associated with a peak. One factor affecting the closeness of fit was the presence or absence of a similar competitor species. The presence of only one species in each feeding niche quantitatively defined by seed measurements is evidence for competitive exclusion. Statistically it is highly improbable that the pattern could have arisen by chance.

Adaptive landscapes are not static; they change when the environment changes. To emphasize change, Merrell (1994) coined the term "adaptive seascape" for an ever-changing configuration of fitness maxima! Not all fitness peaks existed on Galápagos islands when the ancestral species arrived. Peaks increased in number with the arrival of new plants and arthropods. As resources increased, decreased, or changed in proportions, peaks would have increased or decreased in height, shifted in position, been deformed by accretion of new resources to existing peaks, become established in new locations somewhere between existing ones or far from them, split into two (Fig. 10.4), or disappeared altogether. Such dynamics would have created new opportunities for diversification of the finches. Hybridization (chs. 8 and 9) could have contributed to the responses to those opportunities.

There are other implications of this viewpoint. First, some peaks may not be occupied because they are too far from existing peaks, despite the dynamics of peak positions, or because a previous occupant has become extinct. Unrecorded extinction may prevent us from understanding how a species became isolated

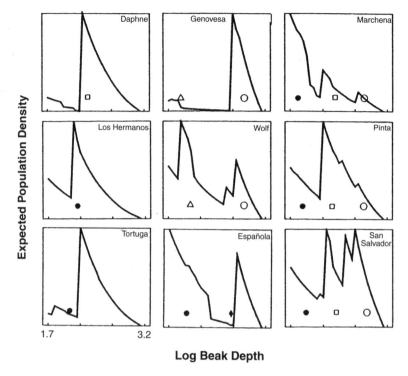

Fig. 10.5 An adaptive landscape for finches, with strong peaks, is revealed by calculating fitness as the expected population density of a solitary seed-eating finch with a particular average beak size (across the range of beak sizes) from the food supply on each island. Actual beak sizes on each island are indicated by symbols for *G. fuliginosa* (solid circle), *G. difficilis* (open triangle), *G. fortis* (open square), *G. conirostris* (solid diamond), and *G. magnirostris* (open circle). Nine examples are shown from the 15 in Schluter and Grant (1984a).

on its peak, such as appears to be the case with warbler finches. Second, the resources that determine the peaks are not entirely independent of their finch consumers: consumers and their resources coevolve. A possible example of coevolution is shown by the particularly large form of *G. magnirostris*, restricted to the only two islands, San Cristóbal and Floreana (Fig. 9.3), where the seeds of one of their food plants, *Opuntia* cactus, are exceptionally large and hard (Grant and Grant 1982). The evolution of large seed size as a result of directional evolutionary pressure exerted by the finches would have fed back on the finches, resulting in a selective favoring of large finches that could crack them.

In theory this evolutionary mechanism could have contributed to peak splitting (Fig. 10.4).

The adaptive landscape was developed by Wright (1932) as a metaphor for the evolution of a population in a heterogeneous environment. Although not intended for this, it can also stand as a metaphor for adaptive radiations that begin when a mountain barrier, valley, or sea is crossed and a substantially new realm of peaks—an archipelago of peaks—becomes accessible.

A Pattern of Ecological Segregation

As species accumulate in an adaptive radiation, competitive interactions are likely to increase and become more intense (Perkins 1903, Simpson 1949) unless the diversity of resources increases through immigration at a faster rate than does the diversity of consumers. Competition between species is minimized by occupation of different habitats, or specialization on different food types or food sizes within a habitat through evolutionary change in beak morphology (Perkins 1903, Lack 1947, Amadon 1950, Pratt 2005). Thus the question arises, is there predictability in the order in which ecological segregation evolves in a radiation? Diamond (1986) concluded from a study of montane birds in New Guinea that habitat separation of closely related species precedes the evolution of dietary differences. Some support has been found from other studies of birds (Richman and Price 1992; also, Richman 1996, Price et al. 2000) as well as lizards (Losos 1994, Losos et al. 1998).

In one respect the Darwin's finch radiation is consistent with this pattern. Species formed early in the radiation occur together less often than later ones do. Warbler finches occur either in moist upland forest (*olivacea* lineage), or mid-elevation and lowland forest (*fusca* lineage). Vegetarian finches occur mainly in transition forest at middle elevations, and the sharp-beaked ground finches occupy *Zanthoxylum* forest above the transition forest (Plate 7). However, species in the early set are separated also by food type and food size, therefore determining which of these segregating modes came first is not possible. Once again, extinction has possibly deprived us of the missing links that would help to answer the question of which came first. As for the late set of species, tree finches are concentrated in the moist highlands where different species co-occur, and ground finches are concentrated in arid lowlands and occur together, but members of the two genera can be found in the same habitat in the breeding season, for example on the top of the highest island, Fernandina. They are ecologically segregated by habitat to some extent, but conspicuously segregated by food type and size. If the ecology of recently formed species reflects the early and now missing stages of diversification, segregation by niche precedes segregation by habitat. This has been found in a recent analysis of some plant species (Ackerly et al. 2006).

SPECIALIZATION

Perkins (1903) developed the idea that the Hawaiian honeycreeper finch radiation proceeded by successive specializations in beak morphology to the foods provided by lobelias and arthropods. Simpson (1949) believed that increasing specialization is one of the most widespread features of evolution, and Lack (1947) applied the idea to Darwin's finches when interpreting the adaptive radiation. In fact there is no clear pattern among the finches, although the subject has not been investigated as fully as possible. On the one hand, the species with the most specialized feeding habits evolved somewhat late in the radiation. The woodpecker finch is the best example. It uses a cactus spine or twig as a tool to extract cryptic arthropod prey (Plate 29). Sharp-beaked ground finches are another example. On the island of Wolf they peck at the developing wing feathers of seabirds and drink the blood (Plate 14). On the other hand, there is a tendency for the early set of species to be restricted to a single habitat more than the later set. Some early specialists may have become extinct; specialists are more vulnerable to ecological disturbance than generalists. Schluter (2000) analyzed several sets of data from different adaptive radiations including Darwin's finches and concluded that there is no quantitative support for the proposed trend. Simpson's statement about "divergence and specialization" would be better recast as divergence and adaptation to new resources.

THE BUILDUP OF COMPLEX COMMUNITIES

The contemporary avian community on Galápagos has been built up from a simpler one by evolutionary generation of diversity of finches and mockingbirds (*Nesomimus* spp.), together with immigration of nine other terrestrial species from the continent. Apart from the yellow warbler (*Dendroica petechia*), the others—dove (*Zenaida galapagoensis*), cuckoo (*Coccyzus minor*), martin (*Progne subis*), hawk (*Buteo galapagoensis*), flycatchers (*Myiarchus magnirostris* and *Pyrocephalus rubinus*), and owls (*Tyto alba* and *Asio flammeus*)—are ecologically very different from finches and mockingbirds, and from each other. This may reflect selective establishment of immigrant species from among those that have reached the Galápagos islands. In other words, food supply, habitat requirements, and competition have guided the development of the whole terrestrial avian community, and not just the Darwin's finch component.

A reduction and simplification of the environment could set this process in reverse. Conversion of forests to agriculture and introduction of alien vertebrates have jointly resulted in the extinction of some populations of finches (Grant et al. 2005b). And now there is deep concern over the possible effects of avian malaria (*Plasmodium*), should it arrive, in the light of experience in Hawaii, where it has had a major impact on the endemic honeycreeper finch fauna (van Riper et al. 1986). The mosquito vector, *Culex quinquifasciatus*, is already present in Galápagos (Whiteman et al. 2005). Perhaps more seriously, *Philornis downsii*, a nest parasitoid, is more than a potential threat, it is a real threat. The larvae of this dipterous fly kill finch nestlings, and have had a large impact on reproduction of finch populations in both highlands and lowlands of large islands (Fessl and Tebbich 2002, Fessl et al. 2006). This shows all too clearly that community buildup in the past depended on the presence of suitable habitat and the absence of enemy organisms.

Summary

In this chapter we attempt to interpret the radiation of Darwin's finches by paying attention to the ecological circumstances in which different speciation cycles took place. The radiation unfolded with an increase in number and diversity of species in a changing environment, and it was molded by natural selection, introgressive hybridization, and extinction. An increase in number of islands increased the opportunities for speciation and thereby the *number* of species. A change in climate and altered vegetation increased the opportunities for *new types* of species to evolve. Species that evolved early in the radiation differ in three ways from those that evolved relatively late: they display (a) almost the complete generic, morphological, and ecological diversity of the whole group, yet (b) no species diversity within genera, and hence (c) no sympatry within genera. These differences can be accounted for either by low speciation rates or by a high rate of extinction of early species. It is not possible to distinguish their separate effects because neither is directly known. Nevertheless, there are reasons to suggest that both processes have been at work: speciation was probably slow in the early stages of the radiation because ecological opportunities were limited, and extinction of some of the early products of the radiation was caused by the combined influence of competitors and changing habitat conditions. There is no clear pattern of increasing specialization, in contrast to an earlier suggestion that increasing specialization is typical of adaptive radiations. A recurring theme of the chapter is that extinction, both anthropogenic and natural, obscures some aspects of the radiation. Extinction, like introgressive hybridization early in the radiation, is invisible history. Allowance for both extinction and introgression alters the interpretation of the early sequence of the radiation given in chapter 2.

CHAPTER ELEVEN

Facilitators of Adaptive Radiation

[I]n my opinion the peculiarities of the finches are primarily due to
an unusual combination of geographical and ecological factors.
(Lack 1947, p. 148)

[I]ntrogression of genes, not adaptive *per se*, can lead to rapid
adaptive evolution.
(Lewontin and Birch 1966, p. 335)

INTRODUCTION

DARWIN'S FINCHES ARE UNIQUE. Apart from mockingbirds (four
allopatric species; Plate 30) other land birds have not diversified in the
Galápagos; in fact only one other group of terrestrial animals, *Bulimulus* snails (Parent and Crespi 2006), has undergone an adaptive radiation
there. Some invertebrates have speciated but not radiated adaptively. Tenebri-
onid beetles (genus *Stomium*) provide an example. An ancestral colonizing
species gave rise to 13 endemic species, all flightless and with a biogeographic
history similar to that of the warbler finches (Finston and Peck 2004, Tonnis
et al. 2004). Morphological differences among them can be interpreted as
adaptive, with difficulty, but founder effects and random genetic drift are
believed to have been the main causes of the rather small amount of differentiation (Finston and Peck 1997, 2004). Sympatry is absent: although three
islands have more than one species they are never found together. Other
insects (orthopteroids) have repeatedly evolved an adaptation (flightlessness)
to the Galápagos environment, but it has not been accompanied by speciation
with establishment of sympatry (Peck 1996).

Lack of time can explain lack of radiation in some cases, but not why tortoises have diversified relatively little, since they have had almost as much time (2 MY; Caccone et al. 1999, Beheregaray et al. 2004) as Darwin's finches. Other organisms have been present in the archipelago for longer than Darwin's finches, and longer even than all extant islands, nevertheless they have not radiated. This shows that a long time by itself does not guarantee an adaptive radiation. Flightless *Galapaganus* weevils (Sequeira et al. 2000), lizards (Lopez et al. 1992), geckos (Wright 1983, Kirzian et al. 2004), and iguanas (Rasmann 1997) started on their respective evolutionary paths more than 5 MY ago on Galápagos islands that are now submerged. Aside from the strikingly different land and marine iguanas these are all non-adaptive radiations (Gittenberger 1991, Kozak et al. 2006, Wake 2006); morphological and genetic divergence, and known or suspected reproductive isolation, have not been accompanied by any ecological divergence that has been recognized so far. Species have multiplied but not conspicuously diversified. The reasons for this are not obvious.

What makes Darwin's finches so special? Answers are to be found in the finches and in their environment. There are two major ideas, extrinsic and intrinsic to the finches: they experienced greater ecological opportunities to diversify than other bird species, and they possessed high speciation potential. The combination made them different from all other birds in Galápagos.

Environmental Opportunity

[I]n my opinion the peculiarities of the finches are primarily due to an unusual combination of geographical and ecological factors. The chief geographical factor has been the existence of a number of islands, which has provided favourable opportunities for the differentiation of forms in isolation and their subsequent meeting. The chief ecological factor has been the scarcity of other passerine birds, which has permitted an unusually great degree of ecological divergence between forms, and thus has allowed an unusually large number of related species to persist alongside each other without competition. (Lack 1947, p. 148)

These two factors facilitate adaptive radiation, although neither is required for it.

Geographical Suitability

In support of Lack's geographical argument, oceanic archipelagoes other than Galápagos possess examples of exceptionally diverse groups of organisms living in a single region. They include Hawaii, with a more diverse radiation of honeycreeper finches (Plate 31) than Darwin's finches (Pratt 2005), spiders (Gillespie 2004), and numerous insects (Perkins 1913, Zimmerman 1948), notably *Drosophila* (DeSalle 1995); and the Caribbean, with a high diversity of *Anolis* lizards (Williams 1972, Losos 1998) and *Eleutherodactylus* frogs (Hedges 1989). Typically islands within archipelagoes are isolated from each other less than they are from a continental source. Speciation cycles are facilitated by isolation, but not too much (or too little).

The endemic vangas on Madagascar (Plate 31) demonstrate that a radiation is possible within a single island providing it is large and heterogeneous (cf. Diamond 1977). At least 15 species, and possibly 19, evolved in 1.5–3.0 MY (Yamagishi and Honda 2005), which is roughly the same number of species as Darwin's finches and produced in roughly the same amount of time (Fig. 10.2). They are part of a broader radiation (Cibois et al. 1999). The vangid starting point, as with Darwin's finches and possibly with Hawaiian honeycreepers (Pratt 2005), was a small bird, conforming to a rule (Cope's rule) that size generally increases during radiations from small beginnings. They have diversified further than the finches, with the extremes being sickle-shaped and helmut-shaped beaks. Like the finches, species with different beak sizes and shapes feed in different ways that can be interpreted adaptively (Yamagishi and Eguchi 1996), and they are often sympatric. The problem posed by the vangids is to explain how the adaptive radiation occurred not in an archipelago but on a single island. The island is topographically varied enough to allow evolution in geographical isolation and this, coupled with climatic fluctuations that have affected habitat distributions, enables investigators to invoke the allopatric model of speciation (Wilmé et al. 2006).

The same type of explanation has been offered for the evolution of the giant, flightless birds (moas), now extinct, on the two islands of New Zealand. Their history has been reconstructed with ancient mtDNA extracted from fossil bones (Baker et al. 2005). Fourteen monophyletic lineages in six genera diversified between 4 and 10 MYA. The question of how an ancestral species of moa could split into several derived species on a single island or two is answered by invoking tectonic activity and mountain building, with climate cooling, that led to geographical isolation and ecological specialization within islands. In similar ways the archipelago-like distribution of mountain tops, lakes, or isolated valleys in continental regions are believed to have been conducive to speciation (Fjeldså and Lovett 1997, Roy 1997). To quote one example, "[F]or continental [African] birds the most active speciation is a time of unfavorable climate, such as will break its natural habitat into small blocks or 'islands' " (Hall and Moreau 1970, p. x).

ECOLOGICAL OPPORTUNITY

David Lack's (1947) principal answer to the question of why Darwin's finches radiated and other land birds did not is that they were the first to arrive in Galápagos. By doing so they gained the twin advantages of a long time in which to diversify and initial freedom from competitors and predators:

> That Darwin's finches are so highly differentiated suggests that they colonized the Galápagos considerably ahead of the other land birds. . . .
> The absence of other land birds has had a most important influence on the evolution of Darwin's finches, since it has allowed them to evolve in directions which otherwise would have been closed to them. . . . After their long flight across the ocean, the ancestors of Darwin's finches entered into a land of abundant foods and varied living quarters, unmarred by the presence of competitive neighbours. This avian paradise probably possessed another considerable amenity, namely, complete freedom from enemies. . . . The absence of predators probably means that Darwin's finches have been limited in numbers primarily by their food supply. When this is the case, adaptations in feeding methods are likely to be of special importance in determining the survival of species, so that the absence of predators may well have accelerated the adaptive radiation of the finches. (Lack 1947, pp. 113–114)

The first-arrival hypothesis can be tested by dating the divergence of other Galápagos land bird species from their continental relatives. The hypothesis survives the test with hawks (< 0.2 MY; Bolmer et al. 2006) and yellow warblers (< 1.0 MY; Browne and Collins MS). However, mockingbirds may have arrived on Galápagos at about the same time as the finches (Arbogast et al. 2006), hence there is serious doubt about the prior arrival of the finches.

Estimates of the time of separation of Galápagos mockingbirds from their mainland/Caribbean relatives are between 1.6 and 5.5 MYA. If the earlier half of this range of dates (3.5–5.5 MYA) is accepted, then the first-arrival hypothesis for the finches is clearly wrong. If the most recent date is accepted, the hypothesis is upheld. However, if this is correct, it means the rate of species accumulation will have been as fast as in Darwin's finches. Equal rates call into question the reasoning that underpinned Lack's first-arrival hypothesis: that the high finch diversity is explained by (longer) elapsed time and not by (faster) rate of diversification.

A simple measure of the rate is the time interval over which the number of species is doubled. It could be 0.75 MY for both groups, yielding four species of mockingbirds in ~1.5 MY and 14 species of finches (16 expected) in ~3.0 MY. Thus there are two possibilities: either Darwin's finches were the first to arrive and diversified at the same rate as the mockingbirds, but for longer; or they arrived at about the same time but diversified twice as fast. Better estimates of times of arrival from molecular data are needed to resolve this issue. The comparable issue has been resolved in Hawaii: the ancestral honeycreeper finches were *not* the first to arrive, they were preceded by honeyeaters (Family Meliphagidae; Fleischer and McIntosh 2001). The pioneering honeyeaters did not prevent some of the honeycreepers from adapting to the pollen-and-nectar niches provided by flowers (Plate 31).

Earlier writers also stressed that adaptive radiations take place when new environments are entered and species that might compete for food are absent or scarce. Radiations proceed by multiple speciations and the filling of all available niches (Perkins 1903, 1913, Huxley 1942, Simpson 1949). In the colorful language of T. H. Huxley, evolution is a process of filling the ecological barrel (Simpson 1949), and stops when the barrel is full. For adaptive radiations this view, in its simplest form, assumes that all resources are present at the outset, as does Lack's ecological hypothesis: an empty barrel is there to be filled. It implies that diversification is rapid and most niche space is filled early in the radiation; then it slows down as ecological opportunities diminish and the remaining gaps in ecological space are filled.

Some, perhaps most, radiations conform to this pattern, for example horses (MacFadden and Hulburt 1988), lizards (Harmon et al. 2003, Vitt and Pianka 2005), warblers (Lovette and Bermingham 1999), and cichlid fishes in the Great Lakes of Africa (Seehausen 2006). The Darwin's finch phylogeny, unknown to Lack, shows exactly the opposite pattern at face value, with an apparently slow start and most rapid diversification recently (ch. 10). They radiated so strongly because they, unlike other animals, diversified when the environment changed, just as the Hawaian honeycreeper finches (Plate 31), but apparently not the honeyeaters (Pratt 2005), radiated strongly when their food supply diversified (Perkins 1903). We see the ability to diversify in response to a changing environment as the key to their evolutionary success. Returning to the metaphor of an ecological barrel, it is best to think of it as not fixed, but capable of expansion through interactions among its occupants (Simpson 1949), and restructured through immigration of new resources. We suspect the same is true of other radiations for which a constant environment is usually assumed for convenience and in the absence of information to the contrary.

HIGH DIVERSIFICATION POTENTIAL

Lack's first-arrival hypothesis assumes that any of the other species could have radiated if they had arrived first; that yellow warblers, for example, would have radiated into many species, though not necessarily 14, had they been the lucky ones to arrive first. Equal potential among all colonists seems highly unlikely. We suggest that Darwin's finches diversified and the other bird species did not because the ancestral finches possessed a much higher intrinsic potential to diversify than the others. Diversification potential is an elusive concept, as elusive as ecological opportunity, because it is recognized only after it has been expressed. It is nonetheless real, as indicated by the radiation of Darwin's finch relatives in another archipelago, the Caribbean (Burns et al. 2002) (Fig. 11.1). There are many possible ecological and reproductive ingredients of higher diversification potential. One is a generalized ancestral beak with omnidirectional possibilities for change, as was the case for Hawaiian honeycreepers (Pratt 2005). For Darwin's finches two other candidates are behavioral flexibility and hybridization.

BEHAVIORAL FLEXIBILITY

Darwin's finches display some unique feeding habits. These include the use of a cactus spine, twig, or leaf petiole as a tool to extract food by woodpecker

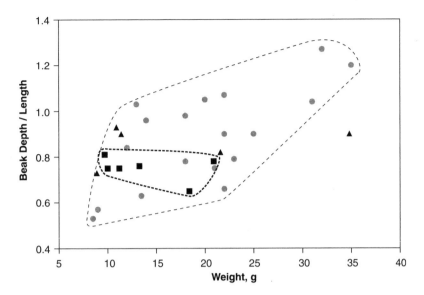

Fig. 11.1 Darwin's finches (outer polygon) radiated much more extensively than their Caribbean and South American relatives over the same period of time, especially in beak proportions. The differences (morphological disparity) among six species (squares) in three genera of the Caribbean birds (*Tiaris, Melanospiza, and Loxigilla*) are shown as a zone nested within the outer broken line enclosing the Darwin's finch points (circles) from Fig. 10.1. Members of the three Caribbean genera that are older than the oldest Darwin's finch species (from Burns et al. 2002) are indicated by triangles. Mean beak measurements are of males only, from Ridgway (1901), Lack (1947), Van Remson (pers. comm.), and Wenfei Tong (pers. comm.), and weights are from Grant et al. (1985) and Dunning (1992).

finches (Millikan and Bowman 1967, Tebbich et al. 2001), possibly also mangrove finches (Curio and Kramer 1964) and warbler finches (Hundley 1963); tick-eating from iguanas (Plate 29) and tortoises; egg-eating; and feeding on the blood of boobies (*Sula*) from wounds that the sharp-beaked ground finch inflicts (ch. 5), or the blood of sea-lion placentae (Grant 1999). We have seen other unusual, possibly unique, behaviors. On Daphne two *G. scandens* were repeatedly observed catching *Microlophus* [*Tropidurus*] lizards by the tail, causing them to break off, and then eating the autotomized, wriggling tail. *G. fortis* has been seen to capture spiders on a web by hauling in the silk thread, using a foot to hold firm each loop (Plate 16), and on Genovesa we have seen *G. conirostris* do the same. We have also seen this species feed on an insect or spider in a rolled leaf by tapping on one end of the retreat and rapidly moving to the other to intercept it in its attempted escape.

These examples indicate an ability to learn new feeding behaviors through trial-and-error learning during the juvenile development of feeding skills (Grant and Grant 1980, Tebbich et al. 2001, Tebbich and Bshary 2004). Young finches peck at rocks, twigs, and bark as well as more rewarding objects in their first month out of the nest, then gradually restrict their activity to food items (Grant and Grant 1980). Copying others occurs but is not essential for learning (Tebbich et al. 2001). In captive experimental situations woodpecker finches display an ability to solve problems used in primate tests (Tebbich and Bshary 2004). We would expect them to have relatively large brains. Avian species with large brains relative to body size have a high propensity to develop novel feeding traits, and apparently as a result tend to be more successful at becoming established in new environments (Sol et al. 2005).

Such behavioral flexibility could be the non-genetic origin of a shift in ecological niche in the allopatric phase of speciation, followed afterwards by genetic and morphological change in response to selection on any heritable character that would increase the efficiency of performing the new task (Price et al. 2003, West Eberhard 2003, Price 2007). This is Waddington's (1953) principle of genetic assimilation. Genetic variation among species is indicated by differences in tool using and exploratory behavior in captivity. Woodpecker finches use tools, whereas cactus finches do so rarely and inefficiently in captivity and other species not at all (Millikan and Bowman 1967). Interestingly, some woodpecker finches in nature appear not to use tools. Whether this represents a cultural difference from the others or a genetic difference in predisposition is not known.

The ability to adjust behaviorally to local feeding conditions may have been an important factor in the diversification. It has two components: development of a new foraging technique, as described above, and deployment of the same technique to solve different problems in different contexts. As an example of the second, the habit of bracing the beak against a rock or ground and kicking backwards is used by sharp-beaked ground finches on Wolf to break the eggs of boobies (Bowman and Billeb 1965), and is used by several and perhaps all ground finch species to kick rocks and thereby expose hidden seeds (De Benedictis 1966).

Introgressive Hybridization

It takes two species to hybridize. None of the other land birds has produced two sympatric species. Darwin's finches have done so repeatedly, and once the first occurrence of sympatry had been realized introgressive hybridization could have been an important factor in their subsequent radiation, in two ways (Grant and Grant 1998a).

First, introgression can replenish the supply of alleles in a population (Fig. 8.8) and thereby maintain high levels of additive genetic variation. We calculated that introgression has effects on quantitative traits like beak size two to three orders of magnitude greater than mutation (Grant and Grant 1994). Unlike mutation, introgression simultaneously affects numerous genetic loci. Not all alleles have beneficial effects in the recipients, hence incorporation into the recipients' genomes is a selective process, resulting in a somewhat mosaic composition of the genomes (Martinsen et al. 2001, Rieseberg et al. 2003, Mallett 2005, Arnold 2006, Patterson et al. 2006). At the extreme, two species that do not interbreed, such as *G. fuliginosa* and *G. scandens* on Daphne Major, may exchange alleles through an intermediary, *G. fortis*, with which both do breed (Grant 1993). Enhancement of genetic composition and variation are important in small finch populations, where variation is likely to be gradually eroded by the combined effects of oscillating directional selection and random genetic drift (Grant and Price 1981, Grant and Grant 1989).

Second, introgression might play a creative role too if a genetically augmented population can more easily respond to selection and evolve along a novel trajectory than would be the case in the absence of introgression (Grant and Grant 1994). This idea stems from a benchmark paper by Lewontin and Birch (1966) on hybridizing fruit flies in Australia. With supporting evidence from the laboratory, they proposed that the flies increased their tolerance of extreme temperatures as a result of introgressed genes, and expanded their geographical range into new habitats not occupied by either of the hybridizing parental species. Variations on this theme have recently been offered to help explain adaptive radiations of fish (Seehausen 2004, Herder et al. 2006). Introgressive hybridization of Darwin's finch species with different allometric relations between two variables such as beak length and beak depth has the potential to weaken the genetic correlation between them, thus making it easier for selection to shift a population in a novel morphological direction relatively

145

unconstrained by genetic correlations with other variables. On the other hand, when the interbreeding species have similar allometries, for example when one species is simply a larger version of another, change in a novel direction can be more strongly constrained, as is the case with *G. fortis* and *G. fuliginosa* (Grant and Grant 1994).

We suggest that episodic introgression (Fig. 9.2) could have played a role in the adaptive radiation of the finches in these two ways. The greatest evolutionary effect of introgression occurs after some genetic difference has arisen between species but before the point is reached when interbreeding incurs a substantial fitness cost (Grant et al. 2004). So far as we know, none of the Darwin's finches has reached this point. All else being equal, the scope for an important role for introgression increases as the number of species increases, therefore it increased as the radiation progressed and is possibly as high now as it has ever been.

Hybridization and Animal Breeding

Introgressive hybridization is analogous to a process used by animal breeders to improve their stock. As Darwin (1859, 1868) was aware, breeders set up separate lines for selection of favored traits and after some period they cross them. Favored traits in previously separate lines are now combined, and fertility that declined due to inbreeding is restored through heterosis or hybrid vigor (Wright 1977, Falconer and Mackay 1995). Sewall Wright built a theory of evolution on the practice and results of animal breeding. His shifting-balance theory, assigning an important role to genetic drift in addition to selection, was extended to explain the origin of reproductive isolation. The idea that genetic drift plays an important part in speciation has been challenged (Coyne et al. 1997). Nevertheless the core elements of divergence by selection in isolation, followed by combination of favored traits through cross-breeding, is theoretically more secure. It is a small step from here to suggest that hybridization of species, followed by backcrossing, can produce new combinations or proportions of traits faster than can be achieved by selection alone in a single isolated population.

Environmental Conditions Conducive to Introgression

In view of the ecological dependence of hybrid fitness (ch. 8), periods of introgression are likely to have been episodic and therefore evolutionarily potent

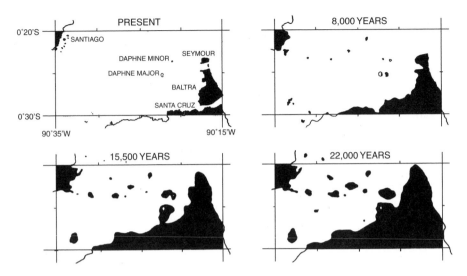

Fig. 11.2 Changes in the number, sizes, and degrees of isolation of islands in the center of the Galápagos archipelago in the last 22,000 years as a result of a lowering of the sea level at the last glacial maximum, followed by a rise in sea level with the melting of polar ice. All dark areas are land. Positions of the two Daphne islands are indicated in white. From Grant and Grant (1998b).

under some conditions and ineffective or entirely absent under others. Small, transitory islands, with unique ecologies and small populations, might have been especially important arenas for introgression. In the last million years or more oscillations between glacial and interglacial conditions at temperate latitudes have caused sea level to rise and fall (ch. 2). Small islands, like small peaks in a dynamic adaptive landscape (Fig. 10.4), have been repeatedly created and eliminated in the Galápagos archipelago (Fig. 11.2) (Grant and Grant 1998b, Grant et al. 2005a). Even on large islands, volcanic activity has decimated and fragmented continuous habitat into small islands, for example on Marchena (Grant 1999). Small populations of finches, although short-lived on a geological time scale, could have been the starting point for species formation through selection and introgressive hybridization. Members of a newly founded population might hybridize with another species owing to a scarcity of mates (Grant and Grant 1997a). The resulting introgression, combined with a loss of original alleles in the bottleneck, could set up the genetic conditions for a new direction of evolution to be taken by natural selection (ch. 4). There is genetic evidence from plants (Rieseberg 1997, Howarth and Baum 2005), insects (Schwarz et al. 2005, Gompert et al. 2006), and corals (Vollmer and Palumbi 2002, Willis et al. 2006) that new species can originate in this way.

An extraordinarily variable population of ground finches on the remote island of Darwin gives a hint of what a stage in this process must have been like. Samples collected a hundred years ago were so variable they were believed to comprise both the large ground finch *G. magnirostris* and the large cactus finch *G. conirostris*. Although they may have been a mixture of *G. magnirostris* residents and *G. conirostris* immigrants, several could have been hybrids and backcrosses with another resident species, the sharp-beaked ground finch *G. difficilis*. A rusty tinge to the plumage of some looks like the telltale sign of *G. difficilis*. Ongoing genetic analysis (with Ken Petren) from fragments of the toe-pads of the museum specimens should help to show whether the cause of enhanced variation was immigration, hybridization, or both.

Finches versus Mockingbirds

We conclude the chapter by comparing finches with mockingbirds because the comparison provides further insight into the Darwin's finch radiation. In contrast to the finches, mockingbirds appear to be an incipient radiation arrested in its early stages. The contrast cannot be explained by absence of competitors and predators, stressed by David Lack, because this factor would apply to both mockingbirds and finches. As mentioned above, the two groups may have been in the archipelago for approximately the same amount of time. If so, we may ask, why are there many ways to be a finch but apparently only one way to be a mockingbird?

Mockingbirds have dispersed to all islands of the archipelago but have failed to establish sympatry on any one of them. Sympatry is also rare on the continent (Arbogast et al. 2006). Assuming secondary contact in Galápagos has occurred in the past, however, there are two possible reasons for this failure: they interbred, or one competitively excluded the other. Both may have happened.

An attempt to interbreed Galápagos mockingbird species in captivity failed (Bowman and Carter 1971). However, an unplanned hybridization occurred in the same experiment, involving a Galápagos mockingbird (*Nesomimus parvulus*) and a mainland species (*Mimus longicaudatus*), which is not the closest continental relative (Arbogast et al. 2006). If mockingbirds so different can produce offspring, surely mockingbird species within the archipelago can do so.

Interbreeding is further suggested by a relative lack of song differentiation among populations (Gulledge 1970). Their learning of songs throughout life (open learning), resulting in a rich repertoire of complex songs (Howard 1974), reduces the chances of song acting as a barrier to interbreeding on secondary contact. Finches are strikingly different. They learn structurally simple songs from their fathers in a short period early in life (closed learning). This facilitates rapid divergence in allopatry and hence establishment of a barrier to interbreeding in sympatry.

We suspect the way in which birds exploit the environment is an additional factor, because a parallel situation exists in the Hawaiian archipelago (Lovette et al. 2002, Pratt 2005). Like the mockingbirds, ecologically similar but unrelated *Myadestes* thrushes have failed to diversify beyond two sympatric species in 4.2 MY (ignoring possible extinctions), whereas more than 50 species of honeycreeper finches evolved (taking account of known extinctions) in as much as 6.4 MY (Fleischer and McIntosh 2001). None of the honeycreeper finches can be described as occupying a thrush niche and preempting further thrush evolution, just as none of Darwin's finches preempted mockingbird evolution. Continental relatives of mockingbirds and thrushes have also not diversified much (Lovette et al. 2002, Arbogast et al. 2006), whereas relatives of Darwin's finches (Burns et al. 2002) and honeycreepers (Lovette et al. 2002) have.

A possible reason for lack of diversification of both mockingbirds and thrushes is their generalist feeding behavior. With their long beaks (Plate 30) they pick up a wide variety of foods from the surface of plants or the ground, probe into leaf litter, and even dig shallowly in the soil. Their diets are broad but constrained by where foods are obtained and their size. Interestingly, Darwin's finches with proportionately long beaks do not coexist with close relatives. The two long-beaked members of the tree finch group (*Camarhynchus* [*Cactospiza*] *pallidus* and *C. heliobates*) occupy different habitats on the same island (Isabela), and the two long-beaked members of the ground finch group (*G. scandens* and *G. conirostris*) do not even occur on the same island. The same applies to the two warbler finch lineages, *olivacea* and *fusca*, which are entirely allopatric. In contrast, tree and ground finch species with blunter beaks that differ in size are fully sympatric. The contrast can be reduced to probing and crushing, and interpreted as; all probers probe alike, but crushers crush differently.

For sustained coexistence without competitive exclusion, perhaps long-beaked members of a feeding guild have to be more different from each other than do blunt-beaked members in environments lacking a rich diversity of flowers. For mockingbirds this may mean they do not diverge enough to permit sustained coexistence unless they first enter new habitats where they are subject to a different set of selection pressures, and then diverge in diet and morphology. Opportunities for segregation of mockingbirds and thrushes by habitat are limited in each of the archipelagoes. A different situation exists in the Antilles, where five species in three endemic genera have evolved in four million years and established sympatry several times (Hunt et al. 2001). A study of their feeding behavior, diets, and food supply could shed valuable light on the ecological requirements for coexistence.

SUMMARY

This chapter offers some explanation for why in the Galápagos setting Darwin's finches are unique in having radiated into at least 14 species, whereas other land birds have produced none, apart from mockingbirds (four allopatric species). There are two major ideas: Darwin's finches experienced greater ecological opportunities to diversify than other species, and they possessed high speciation potential. The ancestral finches may have arrived on the Galápagos first, and by doing so gained the twin advantages of a long time in which to diversify and initial freedom from competitors and predators. Molecular data show that they did arrive before hawks and yellow warblers, but the data for mockingbirds are equivocal. Either mockingbirds arrived earlier and diversification was constrained by their generalist feeding habits, or, if later, they diversified at the same rate as the finches.

We discuss three factors intrinsic to the finches that may help to explain the radiation. The first is their behavioral versatility. Behavioral flexibility could be the non-genetic origin of a shift in ecological niche in the allopatric phase of speciation, followed afterwards by genetic and morphological change in response to selection for increased efficiency in performing the new task. The second is their learned song, which acts as a barrier to interbreeding in sympatry. The third is their proneness to occasionally hybridize. Introgressive hybridization may have been an important factor in the radiation, in two ways. Introgression has the potential to maintain high levels of additive genetic variation in populations, and it might play a creative role too if a genetically augmented population can more easily respond to selection and evolve along a novel trajectory than would be the case in the absence of introgression. Small, transitory islands with unique ecologies and small populations might have been arenas for introgression that contributed to the adaptive radiation by facilitating novel pathways of evolution. The greatest evolutionary effect of introgression occurs after some genetic difference has arisen between species but before the point is reached when interbreeding incurs a substantial fitness cost.

CHAPTER TWELVE

The Life History of Adaptive Radiations

[T]he evolutionary stage shown by Darwin's finches is both
sufficiently advanced to provide a parallel with the more mature
evolution of the continents, and sufficiently early for links to remain
which reveal the underlying processes. . . . The evolutionary picture
presented by Darwin's finches is unusual in some of its details, but
fundamentally it is typical of that which I believe to have taken place
in other birds, while many of the same general principles probably
apply to other groups of animals and to plants.

(Lack 1947, p. 159)

INTRODUCTION

A MAJOR CHALLENGE for evolutionary biologists is to explain the extraordinary richness of biological diversity, from microscopic unicellular organisms to the modern giants of elephants and blue whales (ch. 1). Our response to the challenge has been to investigate the causes of diversification in an adaptive radiation of a single family of birds. We ask why species are as diverse as they are, and why there are so many of them. The answers involve a combination of intrinsic genetic factors that *enable* finches to evolve, and extrinsic ecological factors that both *allow* and *cause* them to evolve and diversity.

Other scientists have asked the same questions of the impressively diverse cichlid fish of African Great lakes (Kocher 2004, Seehausen 2004, 2006), *Anolis* lizards (Losos 1998, Losos et al. 1998) and *Eleutherodactylus* frogs (Hedges 1989) of the Caribbean and Central and South America, and *Drosophila* (DeSalle 1995), honeycreeper finches (Fleischer and McIntosh 2001), and the Silversword alliance of plant species (Barrier et al. 1999) of the Hawaiian archipelago, to mention just a few. To integrate their answers with ours, we use a life history analogy.

Like individual organisms, adaptive radiations have a life history. They are born, they grow, they flourish, they wither, and they die. We are dependent on fossils for a knowledge of their birth and death. For example, without fossils we would not know that marine bivalve mollusks originated (were born) in the tropics (Jablonski et al. 2006), or that the spectacular radiation of dinosaurs and ammonites ended (they died) at the end of the Cretaceous 65 million years ago (Benton 2004, Valentine 2004). Between their beginning and their end adaptive radiations progress through various stages as they grow in size and change in shape through unequal proliferation and extinction of their branches. Again, fossils help to establish the patterns of growth, but contemporary organisms are needed for an elucidation of causes, and can answer some questions that fossils can only raise, such as the role of behavior in speciation.

From the comparative study of adaptive radiations a picture emerges of both unity and diversity. Unity derives from the central role of ecological opportunity in fostering diversification (Schluter 2000), also from introgressive hybridization in enhancing the potential for evolutionary change (Grant and Grant 1998a, Seehausen 2004), especially in young radiations (ch. 11). Diversity derives from different biological properties of the particular groups, from the different environments they occupy, and from their different ages.

These are the themes of a chapter that puts the diversification of Darwin's finches into a broader context by considering what happens in adaptive radiations over much longer periods of time. A comparison of radiations forms a basis for extrapolating to higher taxonomic levels such as classes, phyla, and kingdoms, where evolutionary pathways connecting related taxa are fuzzy at best. Lessons learned from contemporary populations can be further extended to all those extinct organisms like dinosaurs, pterodactyls, ammonites, and trilobites, which collectively far exceed the number of species living in the world today.

The First Stage of Adaptive Radiation

Adaptive radiations, like speciation, pass through recognizable stages. In the first stage, exemplified by Darwin's finches, speciation proceeds with ecological divergence of populations under natural selection in allopatry and the subsequent establishment of sympatry. It involves evolution of pre-mating, but not post-mating, reproductive isolation (Grant and Grant 1997c, Grant 2001). The diversification process happens several times and is relatively rapid. Incipient species compete for food and diverge, but they also hybridize and, depending on the ecological conditions, backcross and converge. Thus the first stage of adaptive radiations is characterized in the sympatric phase of speciation by a tendency to oscillate between fusion due to introgressive hybridization and fission caused by selection. It is a highly creative stage.

The Darwin's finch radiation displays the essence of many features of young radiations. It differs in detail from others as a result of the unique properties of the finches. The finches have dull songs and dull plumage. They do not come close to matching Eurasian warblers (family Sylviidae) in diversity of songs, or North American warblers (family Parulidae) in diversity of plumage colors and patterns (Price et al. 1998, 2000). This implies a greater scope for, and importance of, sexual selection in the radiation of warblers. The same can be said for brightly colored birds such as the birds-of-paradise (Frith and Beehler 1998) and Hawaiian honeycreepers (Pratt 2005), and for the brightly colored cichlid fish of the African Great lakes (Seehausen et al. 1997). On the other hand, cultural evolution through learning has played an important role in the formation of pre-mating barriers between species in the early stages of the radiation of Darwin's finches, and probably of many other passerine birds (Panov 1989, Irwin and Price 1999, Edwards et al. 2005), whereas it has played no known role in the radiation of fish, lizards, or *Drosophila*. Thus radiations proceed through the first stage by different mechanisms, and at different rates—hundreds of cichlid fish species originated in lake Victoria in much less than a million years (Kocher 2004, Seehausen 2006)!

THE SECOND STAGE OF ADAPTIVE RADIATION

The origin of partial but incomplete genetic incompatibility marks the beginning of the second stage of adaptive radiation. At this stage some species, though not all, display intrinsic post-mating isolation to some degree. In a broad survey of birds Price and Bouvier (2002) found that the first signs of reduced fertility of hybrids occurs about 2.5 MY after the interbreeding species diverged from their common ancestor. It takes four times as long for hybrid inviability to begin to evolve. These values for the origin of genetic incompatibilities should be considered minima. Rates were estimated by using a mitochondrial DNA molecular clock, for which an unknown correction factor is needed because an exchange of genes by hybridizing species shortens the genetic distance between them and hence the apparent age of their separation (Fig. 9.2).

The cause of genetic incompatibilities is mutation. Different mutations in the incipient species arise initially in the allopatric phase of speciation, and only have deleterious effects later when, through interbreeding of the species, they become combined in the same genome in hybrid offspring. This is known as the Bateson-Dobzhansky-Muller mechanism after its three independent originators (Orr 1996). Lacking are the details of incompatibilities of genes or chromosomes in individual cases of birds (Price 2007).

Progress through the second stage of adaptive radiation is illustrated in Figure 12.1 with an example from fish; comparable data from birds and other vertebrates have not yet been compiled. This shows hybrid viability declining at an increasing rate, consistent with an accumulation of incompatibilities of small effect (the Bateson-Dobzhansky-Muller mechanism). Notice the remarkably long lag of 10–15 MY before fitness is certain to decline below the fitness of parental species, implicating a period of heterosis (hybrid vigor) in the first stage of adaptive radiation. Darwin's finches (Fig. 8.4) appear to be at an early stage in this lag phase; they are species before speciation is complete (Grant and Grant 2006b). The fish example illustrates the additional point that the rate of evolution of hybrid inviability varies greatly among pairs of species, and that it takes a long time for complete hybrid inviability to evolve.

Partial genetic compatibility also lasts for a remarkably long time in birds. Prager and Wilson (1975) used data on the difference between hybridizing species in their albumin and transferrin molecules (proteins) to show that the potential to produce hybrids decays slowly with time and much slower in birds (and amphibia and reptiles) than in mammals (also Fitzpatrick 2004). With a correction to the calibration of the "transferrin clock" they used (van Tuinen and Hedges 2001), the average persistence of viable hybrid production is 32 MY, and the maximum possible is two and a half times as long! To put the maximum in perspective, it is half the time since *Archaeopteryx* was alive!

Not surprisingly, given its duration among related species and genera, hybridization in birds is widespread (Panov 1989). We extended Panov's survey

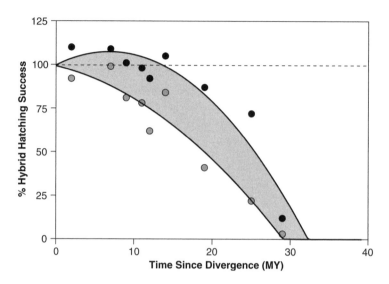

Fig. 12.1 A model of the evolution of genetic incompatibilities between species as a function of time. Although this is based on data from North American centrarchid fishes (sunfish) (Bolnick and Near 2005) it is appropriate for birds and other vertebrates as well. Least successful hybridizations are indicated by grey circles and the lower line (polynomical regression), most successful ones are shown by black circles and the upper line.

and found that approximately one in ten of the nearly 10,000 species of the world are known to hybridize in the wild (Grant and Grant 1992). This is surely an underestimate of the true frequency because so few species have been studied, especially in the tropics. Hybridization is easily overlooked because it is rarely common, and for this reason little is known about the evolution of genetic incompatibilities from field studies that follow the fates of F_1 hybrids and backcrosses (e.g., Rowher et al. 2001, Gill 2004, Curry 2005).

In terms of number of species, an adaptive radiation may reach a peak in stage 2. For birds the prime example of species richness is the Hawaiian honeycreeper finch radiation (Pratt 2005; Plate 31). Fifty species or more evolved from an ancestral cardueline finch that invaded the archipelago 5–6 MYA (Fleischer and McIntosh 2001). The radiation is as old as or somewhat older than the current oldest island (Kaui). Hybridization, and hence genetic compatibilities, is almost unknown (Grant 1994), but this tells us little because more than half the species have become extinct through human activity over the last thousand years or more (Olson and James 1981, James 2004).

HALDANE'S RULE

One important feature of the second stage of radiations is the origin of genetic incompatibilities in a sex-specific manner. As originally formulated by Haldane (1922) the rule is this: if one sex has lower fitness than the other, it is the heterogametic sex. In birds this is the female: the combination of sex chromosomes is ZW. The rule has been attributed to recessive mutations on the Z chromosome that have no adverse effects when the other, non-sex chromosomes (autosomes) are from the same species, but have deleterious effects when combined with autosomes from another species (Turelli and Orr 1995). These adverse effects are counteracted in males (ZZ) by the Z chromosome from the other parent, but not by the W chromosome in females because it has very few loci.

Remarkably, Haldane's rule almost amounts to a law because it is upheld in so many cases. Wu and Davis (1993) assembled 30 cases of crosses between species of birds where one sex in the offspring was infertile; in every case it was the female. Further, in 21 out of 23 cases where only one sex was inviable it was again the female. Price and Bouvier (2002) confirmed the pattern; expectations were met by 72 out of 75 cases of fertility and 15 out of 15 cases of viability. The rule has been shown to apply with similar high consistency to male-heterogametic *Drosophila* and mice (XY), and to female-heterogametic butterflies (Wu and Davis 1993). A particularly high mutation rate at X-linked loci has been demonstrated experimentally (Laurie 1997, Tao et al. 2003).

The best studied example of Haldane's rule in birds is not part of an adaptive radiation, but is applicable nonetheless. European pied flycatchers (*Ficedula hypoleuca*) and collared flycatchers (*F. albicollis*) diverged about 2 MYA (Sætre et al. 2001), or probably earlier given their history of gene exchange. They hybridize at a slightly higher frequency than Darwin's finches, but there is a low rate of introgression because F_1 hybrids are always infertile if female (Alatalo et al. 1990, Tegelström and Gelter 1990) and usually infertile if male (Sætre et al. 2003). The two species differ strongly at Z-linked loci (Borge et al. 2005b).

Nevertheless, introgression of mainly autosomal alleles occurs, more in recent sympatry on Baltic islands than in older sympatry in central Europe, and 50 times more from pied flycatchers to collared flycatchers than vice versa (Sætre et al. 2003, Borge et al. 2005a). Divergence in a plumage trait (reinforcement) may have been caused by selection against mixed matings, for which there is experimental evidence (Sætre et al. 1997), or against males of similar color during territory establishment (Alatalo et al. 1990). Sætre et al. (2003, p. 58) conclude: "We suggest that postzygotic and prezygotic barriers to gene flow may have a common evolutionary origin and that the sex chromosome(s) is the main arena where gene flow is brought to a halt."

THE THIRD STAGE OF ADAPTIVE RADIATION

The second stage of adaptive radiation ends and the third and final stage begins with the complete loss of fertility of hybrids. It is an important point because from then on two species can no longer exchange genes, they become evolutionarily independent, and their fate is determined entirely by the environment. On average this point is reached in birds after about 7 MY (Price and Bouvier 2002), but may take much longer. According to our calculations that employ a transferrin clock (van Tuinen and Hedges 2001), backcrossing of male hybrid pheasants (Gray 1958, Prager and Wilson 1975) still occurs 35 MY after the parental species split apart! This is an extraordinarily protracted loss of fertility, and much greater than in mammals. To the extent that introgression enhances the ability of populations to track environmental change, respond to new ecological opportunities, and avoid extinction, birds retain this potentially valuable source of new genetic variation for a remarkably long time. It is not known why fertility is lost so much more slowly in birds than in mammals (see also Fitzpatrick 2004).

At the same time that complete loss of hybrid fertility evolves, other, generally younger, species in the radiation are still capable of exchanging genes. Introgressive hybridization dwindles in importance as the frequency of hybridization declines with increasing morphological divergence, and as the genetics of incompatibility come to predominate. The 150 species of Caribbean lizards may be an example of this stage, having radiated there in the last 15 MY (Hedges et al. 1992); the degree to which they are genetically compatible is generally not known, however.

The fate of the radiation is determined by the dynamics of speciation and extinction, and they are not likely to be equal. Speciation predominates early in the life of a radiation and extinction predominates later. Furthermore, they are not independent. Paleontologists recognize that extinction and speciation occur closely together in pulses, the former preceding the latter by varying amounts of time (Valentine 1985, Jackson 1994, Vrba et al. 1995). The signs of bursts of speciation can also be recognized in molecular phylogenies. For example, North American *Dendroica* warblers displayed a burst in diversity 4.5–7 MYA, and another 2.3–3.5 MYA (Lovette and Bermingham 1999). Both bursts were associated with conspicuous environmental change: in aridity and the distribution of vegetation (Price et al. 1998, Lovette and Bermingham 1999). They could have been accompanied by elevated extinction rates (ch. 10), but this is conjectural.

With the passage of time in stage 3 the imprint of early history on the radiation fades. An adaptive radiation of closely related species differing in feeding ecology loses its coherence as a result of continuing divergence of some of its members, and extinctions. This is the "thinning out" process of Simpson (1949). Competitive interactions, as stressed by Simpson, may be one cause of extinctions, but there are plenty of other causes involving parasites, pathogens, predators, and changes in habitat and niche resources resulting from climatic change and volcanic activity. Decimated by extinction, an adaptive radiation may no longer be recognizable as an adaptive radiation, particularly if only one species survives! A characteristic feature of old radiations of birds is the presence of only one or two species per genus in related genera. A possible example of a radiation in decline is the non-passerine ground rollers (family Brachypteraciidae) on Madagascar. Known from a fossil to be at least 45 MY old, it is now composed of four species in five genera (Kirchman et al. 2001). Although there may never have been more than five species, there is a good chance that many more existed, forming the missing links between genera.

At the extreme, as the radiation declines it is replaced by another from a different group of organisms. This is revealed repeatedly in the fossil record, often in association with environmental change (Simpson 1949, Valentine 1985, 2004, Gould 2002). Extinctions make habitats available for other groups to exploit. Alternatively a group of animals in one branch of a radiation may re-radiate, evolving and multiplying in a substantially new ecological direction, even as other branches of the same radiation wither and disappear. Birds are an example of re-radiation. They have been called the "feathered dragons" in recognition of their radiation from a branch of the now extinct dinosaurs (Currie et al. 2004). They also show that the initial impetus for a radiation may be a small niche shift, with enormous potential for further evolutionary diversification, as a result of the evolution of a key trait, in their case feathers, which unlocks the door to a new ecological realm.

With the lapse of yet further time, differences among the surviving species and their descendants are large enough to be recognized taxonomically at the family level or higher. In principle there is no limit to the amount of divergence in adaptive radiations that contribute to the evolutionary tree of life. Passerines (order) have radiated within birds (class), birds have radiated within vertebrates (subphylum), and vertebrates have radiated within chordate animals (phylum). Marsupials radiated within mammals, dinosaurs radiated within reptiles (Carroll 1997, Benton 2004). Arthropods radiated spectacularly in the Cambrian more than 500 MYA (Conway Morris 1998), and angiosperm plants have radiated spectacularly in the last 100 MY or so (Labandeira et al. 1994, Soltis et al. 2005). Similar radiations have occurred among microorganisms (Patterson 1999, Travisiano and Rainey 2000, Horner-Devine et al. 2004). They are only less well known and appear less dramatic than the radiation of dinosaurs because their scale is so small. Their numbers and diversity may in fact be huge.

At the bottom of this taxonomic and evolutionary pyramid are radiations like Darwin's finches at an early stage. They are valuable because they help us to understand the top whose origin, details, and causes are lost in the mists of time. Animal phyla are a perfect example, as all of them originated more than a half billion years ago and diverged in the fundamental organization of their body plans. There are no fossils of any of the branching points of origin (Valentine 2004). And yet every pair of closely related phyla was once a pair of closely related species.

Synthesis

Theories are both explanatory and synthesizing devices. They help us to organize knowledge, illuminate areas of ignorance, generate unexpected insights, and identify places where new observations and measurements are needed. Our explanations for how and why Darwin's finches diversified constitute one such theory.

Simpson (1944, 1953) began with fossils, and to explain their diversity he used ecology and genetics of living organisms to construct a theory of adaptive radiation. In contrast, we start with observed ecology, behavior, and microevolution of living birds and use the findings of molecular biology and paleontology to construct a theory of their radiation.

Schluter's (1996, 2000) theory of adaptive radiations, modernizing and extending Simpson's, is explicitly ecological and implicitly genetical. Coyne and Orr's (2004) critical review of genetical theories of speciation is complementary and it illustrates the multiple routes to complete reproductive isolation that organisms can take. While they jointly cover much of what we believe to be important in adaptive radiations in general, they are not sufficient to capture the essential ingredients of how Darwin's finches radiated in particular, or how radiations can be thought of as having a life of their own. For example, an explanation of the finch radiation is incomplete without attention to how pre-mating isolation arises from the learning of cues of potential mates, without consideration of potential genetic effects of hybridization, or without allowance for environmental change. On the other hand an explanation of the Darwin's finch radiation is incomplete as an explanation for other radiations as it omits the evolution of genetic incompatibilities. A more comprehensive, all-inclusive theory is needed. For want of a better word, we use the term synthetic to refer to a theory of adaptive radiation that both captures the essence of Darwin's finch evolution and is applicable to many other organisms as well.

In broad outline the synthetic theory of adaptive radiation is a theory of how species multiply and diversify through natural selection, sexual selection, and chance. It is a theory of how a radiation grows, flourishes, and declines through time as environments change. The main processes are dietary and mate-signal divergence, limited interbreeding with genetic exchange, competition, and repeated mutation that culminates in a genetic barrier to interchange. Previous chapters have described how they occur and why they occur, except for the evolution of genetic barriers (for more details of this, see Coyne and Orr 2004, Price 2007). The present chapter, by adding the genetic barriers, describes the shifting importance of the components during the life of a radiation.

Formal development of the theory in terms of assumptions, hypotheses, and predictions—the stuff of research—is needed to find out how sufficient and general the theory is. The test of a theory is its usefulness as much as its correctness in detail. We hope the synthetic theory of adaptive radiation will help others to understand the diversity of organisms as different from Darwin's finches as bacteria, butterflies, beetles, fungi, and fish.

SUMMARY

In this chapter we put the Darwin's finch radiation in perspective by recognizing three stages of an adaptive radiation. Darwin's finches are an example of the first stage: speciation proceeds with ecological divergence under natural selection in allopatry, establishment of sympatry, and the evolution of pre-mating isolation. This happens several times and is relatively rapid. Incipient species diverge, hybridize, and compete for food, resulting in an oscillation between fusion and fission, or merge and diverge, tendencies. The second stage is dominated by genetic divergence. It starts with the origin of genetic incompatibilities between hybridizing species. The potential to exchange genes through introgressive hybridization diminishes, with fertility of hybrids declining earlier than viability, and female hybrids exhibiting reduced fitness earlier than male hybrids. The third and final stage begins with complete cessation of gene exchange between at least two of the species in a radiation. In birds this point is reached on average after about 7 MY, but can take as long as 30–40 MY. Competition with other groups of organisms continues to be important, and eventually extinction exceeds speciation. The most successful groups of organisms undergo re-radiation into markedly different ecological zones following the evolution of a key trait, when niches become vacant through the demise of their occupants, or when they newly arise through environmental change. In principle there is no limit to the amount of divergence contributed by sequential adaptive radiations to the evolutionary tree of life. We conclude the chapter by proposing a synthetic theory of adaptive radiation in terms of dietary and mate-signal divergence, limited interbreeding with genetic exchange, competition, and repeated mutation that culminates in the formation of a genetic barrier to interchange.

CHAPTER THIRTEEN

Summary of the Darwin's Finch Radiation

WHAT HAPPENED AND WHY

THIS IS WHAT HAPPENED in the adaptive radiation of Darwin's finches
as we currently understand it. A group of finch-like tanagers arrived
on Galápagos from the mainland ~2–3 MYA. The radiation began
when the initial species split into two branches, one giving rise to two lineages
of *Certhidea* warbler finches, and the other giving rise to the Cocos finch, the
sharp-beaked ground finch, and the vegetarian finch, all differing strongly from
each other in their ecological niches. Later products from the second branch
were a group of ground finch species and a group of tree finch species. These
extended the ecological radiation in several dietary directions: they consume
seeds, buds, fruits, arthropods, pollen, nectar, and blood.

To understand why the radiation happened in the way that it did, producing
14 species, we combine contemporary information on microevolution, feeding
ecology, and reproductive biology with estimates of paleoclimates and geophysical history to infer the causes of diversification, as follows. Evolution of
many species derived from a common ancestor was the result of repeated cycles
of speciation. Each cycle comprised an allopatric phase of one or more geographically isolated populations, initiating the speciation process, followed by
an interactive sympatric phase in which it was completed. In the allopatric
phase populations diverged as a result of adaptation to local food conditions,
and to a minor extent through random genetic drift. Beaks, the principal tool
for gathering and dealing with foods, diverged the most in size and shape
under natural selection. Songs also diverged. Songs (acoustic signals) are
learned in association with beak size and shape (visual signals) of parents in an
imprinting-like process during a short sensitive period early in life, and used
later for mate recognition. They diverged partly as a passive consequence of

divergence in body size and beak size, partly through copying errors and other random processes, and partly under the influence of sexual selection. Differences in songs constituted a pre-mating barrier to interbreeding in the sympatric phase. The barrier was not perfect, however, and rare hybridization took place, with the fitness of hybrids and backcrosses being dependent on the external environment and not on internal, post-mating (genetic) factors. Differences in beak size that were acquired in the allopatric phase were enhanced by selection in the sympatric phase, resulting in a reduction of competition for food: a phenomenon known as character displacement. Song divergence (reinforcement) may also have occurred at this time, but evidence is lacking.

The radiation was strongly influenced by two environmental factors: geography and climate change. Galápagos islands are isolated from each other to different degrees and much more strongly from the South American continent; therefore the finches live in an environment too remote to be colonized by many competitor and predator species. The islands vary in elevation and number of habitats and correspondingly they support different combinations of plants and animals. The geophysical and biological environment has not remained constant but changed over the period in which the finch radiation unfolded. An increase in the number of islands provided more opportunities for speciation. Moreover, the Galápagos climate became cooler, probably drier, and changed from permanent El Niño conditions to strongly seasonal and annually fluctuating conditions. Resulting changes in vegetation and associated arthropods increased the opportunities for new types of species to evolve. The overall result of these changes was an increase in the rate of speciation in the last million years, apparently, most likely accompanied by extinction of some of the older species.

Three additional factors facilitated the radiation. First, the absence or scarcity of other species, especially in the early years of the radiation, enabled the finches to diversify relatively unhindered by competitors and predators. Second, flexibility in the early learning of feeding skills gave the species enhanced behavioral potential for exploiting new food resources in new or more efficient ways. Third, introgressive hybridization gave the species enhanced genetic potential for evolutionary change.

Thus a combination of behavioral, ecological, geographical, and genetic factors enabled Darwin's finches to diversify rapidly in a changing, and strongly isolated, environment. These four factors are the A, B, C, and D of speciation and ecological diversification. While all are important, and we emphasize their synthesis and not their separation, behavior plays the key role in birds like

Darwin's finches inasmuch as the origin of new species is the origin of barriers to interbreeding and those barriers are created by behavior.

What Is Missing?

Our current understanding is deficient in at least four areas. First, the history of habitat change on Galápagos is not known from any direct evidence except for the last few thousands of years. Enlightenment on how Galápagos climate and vegetation changed may come from the bottom of the sea: from offshore sediments that have stored a record of the past in the form of plant fragments and spores.

Second, the phylogeny of Darwin's finches is not well enough known to give us confidence in several of the relationships. This applies to affinities with continental and Caribbean species as well as to species within the group such as the ground finches. Use of other molecular markers, such as amplified fragment length polymorphisms (AFLPs; Herder et al. 2006, Seehausen 2006), are likely to improve resolution, perhaps even to the level of populations within species. The causes of the adaptive radiation will be understood better with an improved estimate of phylogeny and a more detailed reconstruction of environmental history, so that the first can be interpreted in terms of the second.

Third, the interbreeding potential of allopatric populations is not known. This cannot be investigated by bringing allopatric populations together because the Galápagos are a National Park, but a possible alternative in the future is artificial insemination experiments in captivity to supplement the playback experiments we have described.

Fourth, we do not know what limits the rate of speciation and the buildup of species-rich communities of finches. In the last few chapters we have followed David Lack in assuming ecological limitation. An alternative that cannot be dismissed is a limiting rate of song divergence and development of barriers to interbreeding. Another possibility is limited dispersal between islands of potentially non-interbreeding and ecologically compatible populations, for example sharp-beaked ground finch populations (ch. 9).

New information is always needed, and old ideas always need to be tested and re-tested. There is more to be learned about the neurobiological basis of song learning and how it varies among species, if at all, how songs diverge, whether a species can split into two reproductively isolated populations on the same island, the long-term consequences of introgressive hybridization, if genetic incompatibilities between species exist, and the reasons for lack of coexistence of well-differentiated allopatric populations. One source of new information is the unplanned experiment of current global warming. It may cause microevolution, and thereby throw light on the influence of persistent climate change on long-term finch evolution.

While some facts appear to be permanently out of reach, new techniques yet to be invented may put them within reach. An era of experimental genome rebuilding in biology lies ahead, and it has the potential of being useful, not only to medical science but to evolutionary biologists (Grant 2000b). With the aid of new techniques of genetic description and engineering (confined to the laboratory) it may be possible to address the following challenges. How many species have become extinct? Did any do so by becoming "reproductively absorbed" into another by hybridization (Grant 1999)? How much morphological change occurred before species became reproductively isolated entities, and how much occurred after? Did the *fusca* lineage of *Certhidea olivacea* give rise to all other species, or is this only apparent and not real, an artifact in the reconstructed phylogeny resulting from introgressive hybridization as we suggested in chapter 10? Did the ancestral species that colonized Galápagos look more like a modern *Tiaris* or a contemporary *Certhidea*? Did more than one species colonize the archipelago and if so did those species interbreed?

An obvious pathway to be followed in the future is the complete characterization of finch genomes. The information to be gained will be more than enough to estimate phylogeny, and rates of mutation and recombination. It will permit reconstruction of evolutionary pathways between species, reveal how variation in beak form arises in development, and perhaps identify the extent and effect of introgressive hybridization in the diversification of the finches. The genetic potential should be coupled with unique opportunities for studying microevolution of finches in the undisturbed environment in which they originally evolved. As the potential benefits are increasingly realized, Darwin's finches will continue to tell us how and why species multiply . . . and diversify.

Epilogue

These are exciting times to be a geneticist: the world of genetics is expanding. At the same time the world's undisturbed environments are shrinking. If we are to take full advantage of genetic discoveries made inside organisms, we need to conserve the environments outside them. Galápagos islands still count among the environments where this is possible. It is our hope this will continue forever.

Forty years ago Dobzhansky (1964, p. 449) famously wrote: "Nothing in biology makes sense except in the light of evolution." For the subject of this book it could also be said: "Nothing in evolutionary biology makes sense except in the light of ecology." This is a second reason for hoping that the light provided by ecology will neither dim nor go out.

Glossary

Adaptive landscape — A conceptual model of the way in which fitness of organisms vary in relation to their traits in a heterogeneous environment.

Adaptive radiation — The rapid evolution from a common ancestor of several species that occupy different ecological niches.

AFLP — Amplified fragment length polymorphisms are small fragments of DNA. When many are available they are used to determine genetic similarities of populations, and similarity values are then used to reconstruct phylogenies.

Albumin — Egg-white protein that has been used in phylogenetic studies. The difference in amino acid composition of the protein between two species is measured by immunological methods, by a technique known as microcomplement fixation. The longer the two species have been isolated from each other the greater is the immunological difference between them. See *molecular clock*.

Alleles — Alternative forms of a gene at a particular locus.

Allometry — The way in which variation in one part of an organism changes in relation to variation in another: for example, the change in beak size in relation to body size.

Allopatry — Different geographical regions occupied by populations.

Allozymes — Different enzymes coded by different alleles at the same genetic locus.

Anthropogenic disturbance — Human-caused disturbance to the environment, often leading to extinctions.

Assortative mating — The mating of similar individuals.

Autosomes — Chromosomes other than the sex chromosomes.

Backcross — Offspring produced by the breeding of an F_1 hybrid with a member of one of the parental species.

Barrier to interbreeding — Any feature of an individual that reduces the chances of its breeding with a member of another population.

Bimodality — Two peaks in a frequency distribution, for example in the distribution of beak sizes in a sample of measurements.

Bootstrap value — Repeated sampling of genetic data used in phylogenetic reconstruction, providing a statistical measure of confidence in the placement and identity of the nodes.

Character displacement — Enhancement of differences between coexisting species by natural selection. The characters that diverge may have ecological or reproductive functions, or both.

Character release — Morphological convergence on a species in its absence; the opposite of character displacement.

Chromosomal rearrangement — A change in the position of a part of a chromosome, for example inverted in reverse order in the same chromosome or translocated into another.

Chromosome — A DNA molecule in the nucleus or mitochondrion where genes reside. Nuclear chromosomes containing factors affecting sex are called sex chromosomes, and the remainder are autosomes. Sex chromosomes are identified by the symbols X and Y in most organisms, and as Z and W in birds.

Class — A category in Linnean classification, above *order* and below *phylum*.

Colonization — Establishment of a new breeding population by immigrants.

Competition — The seeking of a resource in short supply, e.g., food, mates.

Confidence interval — The range of values either side of an estimated mean within which the true mean lies at some specified level of statistical confidence, typically 95 percent.

Dipterocarp — Members of a very successful family of trees, the Dipterocarpaceae, principally in Asia.

Disparity — Degree of morphological differences among a group of species.

Diversification — Evolutionary increase in the number of related species, usually accompanied by an increase in the diversity of phenotypes.

Diversification potential — Intrinsic properties of organisms that facilitate the multiplication of species.

DNA — Deoxyribonucleic acid: the hereditary material of the chromosomes.

Ecological opportunity — Availability of niches in an environment.

El Niño — Occurrence of unusually warm surface waters in the Pacific Ocean, and associated heavy rains.

Electrophoresis — Method of separating and identifying proteins or DNAs by their mobility on a gel in an electric field according to their sizes and electrical changes.

Endemic — Refers to organisms that are restricted to a region.

ENSO — The El Niño–Southern Oscillation; a more general term than El Niño that links the oceanographic phenomenon to an oscillation in atmospheric pressure across the Pacific.

Epithelium — Tissues of cells that cover the outside of the body and organs within the body.

Epistasis — An interaction between genes at different loci producing an effect on phenotype or fitness.

Estradiol — A hormone in the estrogen family that influences female reproduction.

Evolution — Change in a population from one generation to another, in genetic composition (organic evolution) or socially transmitted traits (cultural evolution).

Extinction — Disappearance of a breeding population.

F₁ hybrid — First-generation offspring produced by interbreeding species.

Family — A category in Linnean classification, above *genus* and below *order*.

Fitness — Ability to survive and reproduce, measured by the number of offspring produced.

Foraminifera — Single-celled organisms living in surface seawater.

Founder effect — Change (reduction) in allele frequencies associated with the founding of a new population by a small number of individuals.

Founder event — Establishment of a new population, typically by few individuals.

Frequency dependence — Dependence of the fate of a group of individuals (or alleles) on their frequency in a population.

Frequency modulation — The change in frequencies (pitch) of song notes. The opposite of frequency modulated notes are pure tones.

Gamete — A reproductive cell (egg or sperm).

Gene flow — Addition of genes to a population through interbreeding with members of another population, usually of the same species.

Genetic assimilation — Incorporation of a genetic variant in a population that enhances a trait which is initially not genetic but induced by the environment.

Genetic bottleneck — Reduction in number of alleles associated with a short-term reduction in population size, either in the founding of a new population or subsequently through genetic drift.

Genetic drift — Random change in frequencies of alleles from one generation to the next.

Genetic incompatibility — Reduced fitness of hybrids arising from adverse interactions between maternal and paternal genomes during growth and development.

Genetic recombination — Changing associations among alleles, which occur in two ways. Segments of chromosomes are exchanged by crossing over during meiosis in the formation of gametes, and two parents make new combinations of chromosomes in the offspring.

Genome — The complete set of chromosomes of an individual from one parent.

Genotype — Genetic constitution of an individual.

Genus — A category in Linnean classification, above *species* and below *family*.

Granivore — An eater of seeds.

Haldane's rule — The rule first stated by the geneticist J.B.S. Haldane that when one sex is less fit than the other in hybrids it is the *heterogametic sex*.

Heritability — A measure of family resemblance as a result of shared genes.

Heterosis — Hybrid vigor: the greater average fitness of hybrids compared to the parental species.

Heterogametic sex — The sex with X and Y (or Z and W) chromosomes; the homogametic sex has XX or ZZ chromosomes. In birds females are heterogametic, in humans males are hetorogametic.

170

Heterozygote — Different alleles at the same genetic locus, one on each of the paired chromosomes. Compare *homozygote*.

Heterotypic — A general term meaning different type.

Homozygote — Two copies of the same allele at a locus. Compare *heterozygote*.

Hybridization — Interbreeding. Usually restricted to different species.

Imprinting — Learning early in life during a short sensitive period to respond to particular animals. It leads to attachment to parents (filial imprinting) and to the use of learned information later in life in mate recognition and pair formation (sexual imprinting). Also used for an analogous process of learning features of the environment.

Immigration — Arrival of individuals from another population.

Inbreeding depression — Loss of fitness experienced by the offspring of kin-related parents, often as a result of an accumulation of deleterious recessive alleles in homozygous condition.

Introgression — Transfer of alleles from one species to another as a result of hybridization and backcrossing.

Isolating mechanism — A difference between populations that restricts or prevents gene exchange.

Kingdom — A taxonomic category above the level of phylum, e.g., animal kingdom.

La Niña — The opposite oceanographic condition of El Niño: unusually low sea surface temperatures.

Least squares regression — A statistical method for estimating the relationship between two variables, when one (the predictor) can be used to predict the other. The relationship may be linear, or it may be curvilinear and described by an equation with two or more predictors (polynomial regression).

Lineage — General term for a line of descent in a pedigree, characterizing populations, individuals, or genes.

Linnean hierarchy — A system developed by Linnaeus to classify organisms at different levels in a hierarchy, from species at the bottom to kingdom at the top.

Locus — Location of a gene on a chromosome.

Mhc locus — The major histocompatibility locus. The proteins it produces have immune functions.

Meiosis — The process of cell division in the formation of gametes (sperm and egg) that have half the number of chromosomes of the original cell. Recombination through chromosomal exchange takes place at this time.

Mesenchyme — Tissue of cells formed early in developing embryos. Neural crest–derived mesenchyme cells migrate to various parts of the embryo and give rise to bones of the skull and other structures.

Microsatellites — Short pieces of DNA that usually have no coding function. They are composed of pairs, triplets, or quartets of nucleotides repeated a variable number of times.

Misimprinting — Imprinting on a member of another population.

Mist net — A fine nylon net used to capture birds.

Mitochondrial DNA — Each mitochondrion in the cell has a single circular DNA molecule. There are many mitochondria in each cell.

Molecular clock — Changes in mitochondrial DNA or protein composition due to mutation accumulate at a roughly constant rate over long periods of time. When the rate of change is known, the difference between two related lineages can be used to estimate the elapsed time since they shared a common ancestor. Fossil evidence establishes the rate at which the clock ticks.

Monophyly — A group of species is monophyletic if they all share the same common ancestor, and share it with no other species.

Mutation — Alteration of a chromosome that is heritable.

Natural selection — A difference in survival or reproduction between members of a population due to a trait they possess or the expression of a trait. Some survive or reproduce better than others in a given environment as a result of the particular traits they have. Selection may be directional, in which individuals with traits at one extreme do best; disruptive, in which individuals at both extremes do best; or stabilizing, in which intermediate individuals do best at the expense of those at both extremes.

Nested relationship — A group of related species within another group in a phylogeny.

Neural crest — A band of dorsal cells along the neural tube of developing embryos that will eventually become the central nervous system.

Niche — The part of the environment that is used by organisms.

Node — A junction point in a phylogenetic tree.

Nuclear DNA — DNA that resides in the nucleus of a cell.

Nucleotides — The building blocks of DNA.

Order — A category in Linnean classification, above *family* and below *class*.

Pacific plate — Part of the earth's crust in the eastern Pacific Ocean.

Paleoclimate — Climate in ancient times.

Parapatry — Adjacent geographical areas occupied by populations; contiguous allopatry.

Paraphyly — A group of species is paraphyletic if they share the same common ancestor with other species.

Parasitoid — A parasitic species that eventually kills its host.

Peripatric speciation — Formation of a new species on the periphery of a large geographical range of the parental species.

Phenotype — Structural or functional properties of an organism.

Phoneme — The pattern of sound energy distribution in the notes of a song. The appearance of a note on a sonagram.

Photosynthesis — Conversion by plants of light into stored chemical energy using C3 or C4 alternative biochemical pathways.

Phylogenetic species — A species recognized by its genetic relationship to other populations in a phylogeny.

Phylogeny — A grouping of populations on the basis of their relationships, usually genetic, that reflects the pattern of descent from an ancestor, i.e., their evolutionary history.

Phylum — A category in Linnean classification, above *class* and below *kingdom*.

Plankton — Microscopic organisms living in surface waters of oceans or lakes.

Population — A group of interbreeding individuals and their offspring.

Post-mating isolation — A lack of successful interbreeding between members of two populations owing to a failure in fertilization or development of offspring.

Post-zygotic isolation — A lack of successful interbreeding between members of two populations owing to a breakdown in the development of offspring at some time after zygotes are formed; offspring are inviable or infertile.

Pre-mating isolation — A lack of interbreeding between members of two populations owing to differences in mate choice or timing of breeding.

Pre-zygotic isolation — A lack of interbreeding between members of two populations owing to an absence of mating, or mating but a lack of fertilization.

Primordium — Tissue prior to becoming differentiated at that location.

Principal Components Analysis — A statistical method of analyzing combinations of variables, for example beak length, depth, and width, to produce a synthetic measure of beak size or beak shape.

Radioisotope — The radioactive form of a chemical element such as carbon (C^{13}) or oxygen (O^{18}). Decay at a constant rate to the non-radioactive form of carbon can be used to date biological materials. Stable oxygen isotope ratios of foraminifera reflect the sea temperatures they experienced when alive.

Recessive allele — An allele that is only expressed in homozygous condition.

Reinforcement — Divergence of courtship signals and responses in sympatry as a result of selection; the initial differences are increased.

Reproductive barrier — A function of organisms that hinders or prevents breeding with another.

Reproductive isolation — Lack or scarcity of interbreeding between populations.

Retroviral vector — An RNA virus that carries an experimentally inserted gene and is used by the investigator to express that gene in a chosen tissue. The retrovirus causes the synthesis of DNA complementary to its RNA genome on infection.

RNA — Ribonucleic acid; a family of molecules with a variety of functions in the cell, including the control of protein synthesis.

Sexual imprinting — Early experience of parents affecting mate choice of offspring later in their life.

Sexual selection — A difference in average success in obtaining a mate or mates between individuals with a particular phenotype and those with other phenotypes. The phenotypic traits may be deployed in encounters between members of the same sex (intra-sexual selection) or opposite sex (inter-sexual selection).

Signaling molecule — Molecule that activates others in biochemical pathways within cells and in cell-cell communication.

Sister species — Two species that are more related to each other than either is to another.

Sonagram — Representation on paper of song characteristics (frequency or amplitude) as a function of time.

Speciation — The process of species formation; the evolution of reproductive isolation.

Species — A population or group of related populations that differ from other populations by certain criteria. The criteria include degree of genetic difference, whether the populations interbreed or not, and the consequences of interbreeding. These criteria are debated.

Stabilizing selection — Maintenance of an unchanging average phenotypic trait in a population as a result of intermediate individuals having the highest fitness; see *natural selection*.

Standard deviation — A statistical measure of the degree to which measurements in a sample are dispersed either side of the mean.

Standard error of the mean — A statistical measure of the uncertainty in the estimate of the mean.

Sympatry — A geographical region jointly occupied by two or more populations.

Taxon — The most general term for a unit of classification, such as population or species. The plural is taxa.

Taxonomy — The science and practice of biological classification.

Transferrin — A protein that functions in regulating the iron content of cells. It has been used in phylogenetic studies because the genetic determinants of the protein undergo change through time at an approximately constant rate. See *molecular clock*.

Trill rate — The rate at which individual song notes are repeated.

Zygote — A fertilized egg.

References

Abbott, I., L. K. Abbott, and P. R. Grant. 1977. Comparative ecology of Galápagos ground finches (*Geospiza* Gould): evaluation of the importance of floristic diversity and interspecific competition. *Ecol. Monogr.* 47: 151–184.

Abzhanov, A., W. P. Kuo, C. Hartmann, B. R. Grant, P. R. Grant, and C. J. Tabin. 2006. The calmodulin pathway and evolution of elongated beak morphology in Darwin's finches. *Nature* 442: 563–567.

Abzhanov, A., M. Protas, P. R. Grant, B. R. Grant, and C. J. Tabin. 2004. Bmp4 and morphological variation of beaks in Darwin's finches. *Science* 305: 1462–1465.

Abzhanov, A., and C. J. Tabin. 2004. *Shh* and *Fgf8* act synergistically to drive cartilage outgrowth during cranial development. *Developmental Biology* 274: 134–148.

Ackerly, D. D., D. W. Schwilk, and C. O. Webb. 2006. Niche evolution and adaptive radiation: testing the order of trait divergence. *Ecology* 87 supplement: S50–S67.

Alatalo, R. V., D. Eriksson, L. Gustafsson, and A. Lundberg. 1990. Hybridization between pied and collared flycatchers: sexual selection and speciation theory. *J. Evol. Biol.* 3: 375–389.

Amadon, D. 1950. The Hawaiian honeycreepers (Aves, Drepaniidae). *Bull. Amer. Mus. Nat. Hist.* 95: 151–262.

Anderson, E. 1948. Hybridization of the habitat. *Evolution* 2: 1–9.

Arbogast, B. S., S. V. Drovetski, R. L. Curry, P. T. Boag, G. Seutin, P. R. Grant, B. R. Grant, and D. J. Anderson. 2006. The origin and diversification of Galápagos mockingbirds. *Evolution* 60: 370–382.

Arnold, M. L. 1997. *Natural hybridization and evolution.* Oxford University Press, Oxford, U.K.

Arnold, M. L. 2006. *Evolution through genetic exchange.* Oxford University Press. Oxford, U.K.

Ashton, P. S. 1982. Dipterocarpaceae. *Flora Melanesia Series 1: Spermatophyta (flowering plants)* 9: 251–552. Martinus Nijhoff, The Hague, Netherlands.

Bachtrog, D., K. Thornton, A. Clark, and P. Andolfatto. 2006. Extensive introgression of mitochondrial DNA relative to nuclear genes in the *Drosophila yakuba* species group. *Evolution* 60: 292–302.

Baker, A. J., L. J. Huynen, O. Haddrath, C. D. Millar, and D. M. Lambert. 2005. Reconstructing the tempo and mode of evolution in an extinct clade of birds with ancient DNA: the giant moas of New Zealand. *Proc. Natl. Acad. Sci. USA* 102: 8257–8262.

Baker, A. J., and P. F. Jenkins. 1987. Founder effect and cultural evolution of songs in an isolated population of chaffinches, *Fringilla coelebs*, in the Chatham Islands. *Anim. Behav.* 35: 1793–1803.

Baker, A. J., and A. Mooerd. 1987. Rapid genetic differentiation and founder effect in colonizing populations of common mynas (*Acridotheres tristis*). *Evolution* 41: 525–538.

Baker, P. A., C. A. Rigsby, G. O. Seltzer, S. C. Fritz, T. K. Lowenstein, N. P. Bacher, and C. Veliz. 2001. Tropical climate changes at millennial and orbital timescales on the Bolivian Altiplano. *Nature* 409: 698–701.

Bard, E., B. Hamelin, R. G. Fairbanks, and A. Zindler. 1990. Calibration of the ^{14}C timescale over the past 30,000 years using mass spectrometric U-Th ages from Barbados corals. *Nature* 345: 405–410.

Barker, F. K., A. Cibois, P. Schickler, J. Feinstein, and J. Cracraft. 2004. Phylogeny and diversification of the largest avian radiation. *Proc. Natl. Acad. Sci. USA* 101: 11040–11045.

Barluenga, M., K. N. Stölting, W. Salzburger, M. Muschick, and A. Meyer. 2006. Sympatric speciation in Nicaraguan crater lake fish. *Nature* 439: 719–723.

Barrier, M., B. G. Baldwin, R. H. Robichaux, and M. D. Purugganan. 1999. Interspecific hybrid ancestry of a plant adaptive radiation: allopolyploidy of the Hawaiian Silversword Alliance (Asteraceae) inferred from floral homeotic gene duplications. *Mol. Biol. Evol.* 16: 1105–1113.

Barton, N. H. 1996. Natural selection and random genetic drift as causes of evolution on islands. *Philos. Trans. R. Soc. B* 351: 785–795.

Barton, N. H. 2001. The role of hybridisation in evolution. *Mol. Ecol.* 10: 551–568.

Barton, N. H., and B. Charlesworth. 1984. Genetic revolutions, founder effects, and speciation. *Annu. Rev. Ecol. Syst.* 15: 133–164.

Barton, N. H., and G. M. Hewitt. 1985. Adaptation, speciation, and hybrid zones. *Annu. Rev. Ecol. Syst.* 16: 113–148.

Beheregaray, L. B., J. P. Gibbs, N. Havill, T. H. Fritts, J. R. Powell, and A. Caccone. 2004. Giant tortoises are not so slow: Rapid diversification and biogeographic consensus in the Galápagos. *Proc. Natl. Acad. Sci. USA* 101: 6514–6519.

Benton, M. J. 2004. *Vertebrate paleontology*, 3rd ed. Blackwells, Oxford, U.K.

Bischoff, H.-J., and N. Clayton. 1991. Stabilization of sexual preferences by sexual experience in male zebra finches *Taeniopygia guttata castanotis*. *Behaviour* 118: 144–155.

Boag, P. T. 1981. Morphological variation in the Darwin's finches (Geospizinae) of Daphne Major Island, Galápagos. Unpubl. Ph.D. thesis, McGill University, Montreal, Canada.

Boag, P. T. 1983. The heritability of external morphology in Darwin's ground finches (*Geospiza*) on Isla Daphne Major, Galápagos. *Evolution* 37: 877–894.

Boag, P. T., and P. R. Grant. 1981. Intense natural selection in a population of Darwin's finches (*Geospizinae*) in the Galápagos. *Science* 214: 82–85.

Boag, P. T., and P. R. Grant. 1984a. The classical case of character release: Darwin's finches (*Geospiza*) on Isla Daphne Major, Galápagos. *Biol. J. Linn. Soc.* 22: 243–287.

Boag, P. T., and P. R. Grant. 1984b. Darwin's Finches (*Geospiza*) on Isla Daphne Major, Galápagos: breeding and feeding ecology in a climatically variable environment. *Ecol. Monogr.* 54: 463–489.

Bolmer, J. L., R. T. Kimball, N. K. Whiteman, J. H. Sarasola, and P. G. Parker. 2006. Phylogeography of the Galápagos Hawk: a recent arrival to the Galápagos Islands. *Mol. Phylogenet. Evol.* 39: 237–247.

Bolnick, D. I., and T. J. Near. 2005. Tempo of hybrid inviability in Centrarchid fishes (Teleostei: Centrarchidae). *Evolution* 59: 1754–1767.

Borge, T., K. Lindross, P. Nádvornik, A.-C. Syvänen and G.-P. Sætre. 2005a. Amount of introgression in flycatcher hybrid zones reflects regional differences in pre- and post-zygotic barriers to gene exchange. *J. Evol. Biol.* 18: 1416–1424.

Borge, T., M. T. Webster, G. Andersson, and G.-P. Sætre. 2005b. Contrasting patterns of polymorphism and divergence on the Z chromosome and autosomes in two *Ficedula* flycatcher species. *Genetics* 171: 1861–1873.

Bowman, R. I. 1961. Morphological differentiation and adaptation in the Galápagos finches. *Univ. Calif. Publs. Zool.* 58: 1–302.

Bowman, R. I. 1963. Evolutionary patterns in Darwin's finches. *Occas. Pap. Calif. Acad. Sci.* 44: 107–140.

Bowman, R. I. 1979. Adaptive morphology of song dialects in Darwin's finches. *J. Ornithol.* 120: 353–389.

Bowman, R. I. 1983. The evolution of song in Darwin's finches. In R. I. Bowman, M. Berson, and A. E. Leviton, eds., *Patterns of evolution in Galápagos organisms*, 237–537. American Association for the Advancement of Science, Pacific Division, San Francisco, CA.

Bowman, R. I., and S. I. Billeb. 1965. Blood-eating in a Galápagos finch. *Living Bird* 4: 29–44.

Bowman, R. I., and A. Carter. 1971. Egg-pecking behavior in Galapagos mocking-birds. *Living Bird* 10: 243–270.

Brazner, J. C., and W. J. Etges. 1993. Pre-mating isolation is determined by larval rearing substrates in cactophilic *Drosophila mojavensis*. II. Effects of larval substrates on time to copulation and mating propensity. *Evol. Ecol.* 7: 605–624.

Bronson, C. L., T. C. Grubb Jr., G. D. Sattler, and M. H. Braun. 2005. Reproductive success across the Black-capped Chickadee (*Poecile atricapillus*) and Carolina Chickadee (*P. carolinensis*) hybrid zone in Ohio. *Auk* 122: 759–772.

Brown, W. L., Jr., and E. O. Wilson. 1956. Character displacement. *Syst. Zool.* 5: 49–64.

Browne, R. A., D. J. Anderson, M. D. White, and M. A. Johnson. 2003. Evidence for low genetic divergence among Galápagos *Opuntia* cactus species. *Noticias de Galápagos* no. 62: 11–15.

Browne, R. A., and E. I. Collins. MS. Genetic variation of yellow warblers, *Dendroica petechia*, in the Galápagos Archipelago.

Buch, L. von. 1825. *Physicalische Beschreibung der canarische Inseln*. Königliche Akademie Wissenschaften, Berlin, Germany.

Bürger, R., and K. A. Schneider. 2006. Intraspecific competitive divergence and convergence under assortative mating. *Amer. Nat.* 167: 190–205.

Bürger, R., K. A. Schneider, and M. Willensdorfer. 2006. The conditions for speciation through intraspecific competition. *Evolution* 60: 2185–2206.

Burkhardt, F., D. M. Porter, S. A. Dean, S. Evans, S. Innes, A. Pearn, A. Sclater, and P. White, eds. 2005. *The correspondence of Charles Darwin: volume 15, 1867*. Cambridge University Press, Cambridge, U.K.

Burns, K. J., S. J. Hackett, and N. K. Klein. 2002. Phylogenetic relationships and morphological diversity in Darwin's finches and their relatives. *Evolution* 56: 1240–1256.

Caccone, A., J. P. Gibbs, V. Ketmaier, E. Suatoni, and J. R. Powell. 1999. Origin and evolutionary relationships of giant Galápagos tortoises. *Proc. Natl. Acad. Sci. USA* 96: 13223–13228.

Cade, T. J. 1983. Hybridization and gene exchange among birds in relation to conservation. In C. M. Schonewald-Cox, B. MacBryde, and W. L. Thomas, eds., *Genetics and conservation*, 288–309. Benjamin/Cummings, Menlo Park, CA.

Cane, M. A., and P. Molnar. 2001. Closing of the Indonesian seaway as a precursor to east African aridification around 3–4 million years ago. *Nature* 411: 157–162.

Carroll, R. L. 1997. *Patterns and processes of vertebrate evolution*. Cambridge University Press, Cambridge, U.K.

Carson, H. L. 1968. The population flush and its genetic consequences. In R. C. Lewontin, ed. *Population biology and evolution*, 123–137. Syracuse University Press, Syracuse, NY.

Carson, H. L., and A. R. Templeton. 1984. Genetic revolutions in relation to speciation phenomena: the founding of new populations. *Annu. Rev. Ecol. Syst.* 15: 97–131.

Chavez, F. P., J. Ryan, S. E. Lluck-Cota, and M. Ñiguen. 2003. From anchovies to sardines and back: multidecadal change in the Pacific Ocean. *Science* 299: 217–221.

Christie, D. M., R. A. Duncan, A. R. Birney, M. A. Richards, W. M. White, K. S. Harpp, and C. B. Fox. 1992. Drowned islands downstream from the Galapagos hotspot imply extended speciation times. *Nature* 355: 246–248.

Cibois, A., E. Pasquet, and T. S. Schulenberg. 1999. Molecular systematics of the Malagasy babblers (Passeriformes: Timaliidae) and warblers (Passeriformes: Sylviidae), based on cytochrome *b* and 16S rRNA sequences. *Mol. Phylogenet. Evol.* 13: 581–595.

Clayton, N. 1988. Song tutor choice in zebra finches and Bengalese finches: the relative importance of visual and vocal cues. *Behaviour* 104: 281–299.

Clayton, N. S. 1990. Subspecies recognition and song learning in zebra finches. *Anim. Behav.* 40: 1009–1017.

Clegg, S. M., S. M. Degnan, J. Kikkawa, C. Moritz, A. Estoup, and I.P.F. Owens. 2002a. Genetic consequences of successive founder events by an island colonizing bird. *Proc. Natl. Acad. Sci. USA* 99: 8127–8132.

Clegg, S. M., S. M. Degnan, C. Moritz, A. Estoup, J. Kikkawa, and I.P.F. Owens. 2002b. Microevolution in island forms: the roles of drift and directional selection in morphological divergence of a passerine bird. *Evolution* 56: 2090–2099.

Colinvaux, P. A. 1972. Climate and the Galápagos Islands. *Nature* 240: 17–20.

Colinvaux, P. A. 1984. The Galápagos climate: present and past. In R. Perry, ed., *Galápagos*, 55–69. Pergamon Press, Oxford, U.K.

Colinvaux, P. A. 1996. Quaternary environmental history and forest diversity in the Neotropics. In J. B. C. Jackson, A. F. Budd, and A. G. Coates, eds., *Evolution and environment in tropical America*, 359–405. University of Chicago Press, Chicago, IL.

Conway Morris, S. 1998. *The crucible of life*. Oxford University Press, Oxford, U.K.

Coyne, J. A., N. H. Barton, and M. Turelli. 1997. Perspective: a critique of Sewall Wright's shifting balance theory of evolution. *Evolution* 51: 643–671.

Coyne, J. A., and A. R. Orr. 1989. Patterns of speciation in *Drosophila*. *Evolution* 43: 362–381.

Coyne, J. A., and A. R. Orr. 1997. "Patterns of speciation in *Drosophila*" revisited. *Evolution* 51: 295–303.

Coyne, J. A., and A. R. Orr. 2004. *Speciation*. Sinauer, Sunderland, MA.

Coyne, J. A., and T. D. Price. 2000. Little evidence for sympatric speciation in island birds. Evolution 54: 2166–2171.

Cracraft, J. 2002. The seven great questions of systematic biology: an essential foundation for conservation and the sustainable use of biodiversity. *Ann. Missouri Bot. Gard.* 89: 127–144.

Crawford, A. J., and E. N. Smith. 2005. Cenozoic biogeography and evolution in direct-developing frogs of Central America (Leptodactylidae: *Eleutherodactylus*) as inferred from a phylogenetic analysis of nuclear and mitochondrial genes. *Mol. Phylogenet. Evol.* 35: 536–555.

Cronin, T. M., and H. J. Dowsett. 1996. Biotic and oceanographic responses to the Pliocene closing of the Central American Isthmus. In J.B.C. Jackson, A. F. Budd, and A. G. Coates, eds., *Evolution and environment in tropical America*, 76–104. University of Chicago Press, Chicago, IL.

Curio, E., and P. Kramer. 1964. Vom Mangrovefinken (*Cactospiza heliobates*). *Zeits. Tierpsychol.* 21: 223–234.

Currie, P. J., E. B. Koppelhus, M. A. Shugar, and J. L. Wright, eds. 2004. *Feathered dragons: studies on the transition from dinosaurs to birds*. Indiana University Press, Bloomington, IN.

Curry, R. L. 2005. Hybridization in chickadees: much to learn from familiar birds. *Auk* 122: 747–758.

Darwin, C. R. 1839. *Journal of researches into the natural history and geology of the countries visited during the voyage of H.M.S. Beagle round the world*. Henry Colburn, London, U.K.

Darwin, C. R. 1842. *Journal of researches into the geology and natural history of the various countries visited during the voyage of H.M.S. Beagle, under the command of Captain FitzRoy, R. N., from 1832 to 1836*. Henry Colburn, London, U.K.

Darwin, C. R. 1859. *On the origin of species by means of natural selection*. J. Murray, London, U.K.

Darwin, C. R. 1868. *The variation of animals and plants under domestication*. J. Murray, London, U.K.

Darwin, F., ed. 1887. *The life and letters of Charles Darwin. Including an autobiographical chapter. Volumes II, III*. J. Murray, London, U.K.

Darwin, F., and A. C. Seward. 1903. *More letters of Charles Darwin. A record of his work in a series of hitherto unpublished letters. Vol. I*. J. Murray, London, U.K.

Davis, M. B., R. G. Shaw, and J. R. Etterson. 2005. Evolutionary responses to changing climate. *Ecology* 86: 1704–1714.

De Benedictis, P. 1966. The bill-brace feeding behavior of the Galapagos finch *Geospiza conirostris*. *Condor* 68: 206–208.

deMenocal, P. B. 1995. Plio-Pleistocene African climate. *Science* 270: 53–59.

de Queiroz, K. 1998. The general lineage concept of species, species criteria, and the process of speciation. In D. J. Howard and S. H. Berlocher, eds., *Endless forms: Species and speciation*, 57–74. Oxford University Press, Oxford, U.K.

DeSalle, R. 1995. Molecular approaches to biogeographic analysis of Hawaiian Drosophilidae. In W. W. Wagner and V. A. Funk, eds., *Hawaiian biogeography: evolution on a hotspot*, 72–89. Smithsonian Institution Press, Washington, DC.

Diamond, J. M. 1977. Continental and insular speciation in Pacific island birds. *Syst. Zool.* 26: 263–268.

Diamond, J. M. 1986. Evolution of ecological segregation in the New Guinea montane avifauna. In J. M. Diamond and T. J. Case, eds., *Community Ecology*, 98–125. Harper and Row, New York, NY.

Dobzhansky, T. 1935. A critique of the species concept in biology. *Philos. Sci.* 2: 344–355.

Dobzhansky, T. 1937. *Genetics and the origin of species*. Columbia University Press, New York, NY.

Dobzhansky, T. 1941. *Genetics and the Origin of Species*, 2nd ed. Columbia University Press, New York, NY.

Dobzhansky, T. 1964. Biology, molecular and organismal. *Amer. Zool.* 4: 443–452.

Dodd, J. R., and R. J. Stanton, Jr. 1990. *Paleoecology, concepts and applications*, 2nd ed. Wiley, New York, NY.

Doebeli, M. 1996. A quantitative genetic competition model for sympatric speciation. *J. Evol. Biol.* 9: 893–909.

Doebeli, M., and U. Dieckmann. 2000. Evolutionary branching and sympatric speciation caused by different types of ecological interaction. *Amer. Nat.* 156 (suppl.): S77–S101.

Dunning, J. B., Jr., ed. 1992. *CRC handbook of avian body masses*. CRC Press, Boca Raton, FL.

Edwards, S. V., S. B. Kingan, J. D. Calkins, C. N. Balakrishnan, W. B. Jennings, W. J. Swan and M. D. Sorenson. 2005. Speciation in birds: genes, geography, and sexual selection. *Proc. Natl. Acad. Sci. USA* 102: 6550–6557.

Endler, J. A. 1977. *Geographic variation, speciation, and clines*. Princeton University Press, Princeton, NJ.

Endler, J. A. 1986. *Natural selection in the wild*. Princeton University Press, Princeton, NJ.

Etges, W. J. 1998. Pre-mating isolation is determined by larval rearing substrates in cactophilic *Drosophila mojavensis*. IV. Correlated responses in behavioral isolation to artificial selection on a life history trait. *Amer. Nat.* 152: 129–144.

Falconer, D. S., and T.F.C. Mackay. 1995. *Introduction to quantitative genetics*. Longman, Harlow, U.K.

Feder, J. L. 1998. The apple maggot fly, *Rhagoletis pomonella*: flies in the face of conventional wisdom about speciation? In D. J. Howard and S. H. Berlocher, eds., *Endless forms: Species and speciation*, 130–144. Oxford University, New York, NY.

Federov, A. V., P. S. Dekens, M. McCarthy, A. C. Ravelo, P. B. deMenocal, M. Barreiro, R. C. Pacanowski, and S. G. Philander. 2006. The Pliocene paradox (mechanisms for a permanent El Niño). *Science* 312: 1485–1489.

Fessl, B., S. Kleindorfer, and S. Tebbich. 2006. An experimental study on the effects of an introduced parasite in Darwin's finches. *Biol. Conserv.* 127: 55–61.

Fessl, B., and S. Tebbich. 2002. *Philornis downsi*—a recently discovered parasite on the Galápagos archipelago—a threat for Darwin's finches? *Ibis* 144: 445–451.

Finston, T. L., and S. B. Peck. 1997. Genetic differentiation and speciation in *Stomium* (Coleoptera: Tenebrionidae): flightless beetles of the Galápagos Islands. *Biol. J. Linn. Soc.* 61: 183–200.

Finston, T. L., and S. B. Peck. 2004. Speciation in Darwin's darklings: taxonomy and evolution of *Stomium* beetles in the Galápagos Islands (Insecta: Coleoptera: Tenebrionidae). *Zool. J. Linn. Soc.* 141: 135–152.

Fisher, R. A. 1930. *The genetical theory of natural selection*. Clarendon, Oxford, U.K.

Fitzpatrick, B. M. 2004. Rates of evolution of hybrid inviability in birds and mammals. Evolution 58: 1865–1870.

Fjeldså, J., and J. C. Lovett. 1997. Geographical patterns of old and young species in African forest biota: the significance of specific montane areas as evolutionary centers. *Biodiversity and Conservation* 6: 325–346.

Fleischer, R. C., S. Conant, and M. P. Morin. 1991. Genetic variation in native and translocated populations of the Laysan finch (*Telespiza cantans*). *Heredity* 66: 125–130.

Fleischer, R. C., and C. E. McIntosh. 2001. Molecular systematics and biogeography of the Hawaiian avifauna. In J. M. Scott, S. Conant, and C. van Riper III, eds., *Evolution, ecology, conservation, and management of Hawaiian Birds: a vanishing avifauna*, 51–60. Allen Press, Lawrence, KS.

Foote, M. 1997. The evolution of morphological diversity. *Annu. Rev. Ecol. Syst.* 28: 129–152.

Ford, H. A., D. T. Parkin, and A. W. Ewing. 1973. Divergence and evolution in Darwin's finches. *Biol. J. Linn. Soc.* 5: 289–295.

Freeland, J. R., and P. T. Boag. 1999a. Phylogenetics of Darwin's finches: paraphyly in the tree–finches and two divergent lineages in the warbler finch. *Auk* 116: 577–588.

Freeland, J. R., and P. T. Boag. 1999b. The mitochondrial and genetic homogeneity of the phenotypically diverse Darwin's ground finches. *Evolution* 53: 1553–1563.

Frith, C. B., and B. M. Beehler. 1998. *The birds of paradise*. Oxford University Press, Oxford, UK.

Futuyma, D. J. 1998. *Evolutionary biology*, 3rd ed. Sinauer, Sunderland, MA.

Garcia–Moreno, J. 2004. Is there a universal mtDNA clock for birds? *J. Avian Biol.* 35: 465–468.

Gavrilets, S. 2004. *Fitness landscapes and the origin of species*. Princeton University Press, Princeton, NJ.

Gavrilets, S., and C.R.B. Boake. 1998. On the evolution of premating isolation after a founder event. *Amer. Nat.* 152: 706–716.

Gavrilets, S., and A. Hastings. 1996. Founder effect speciation: a theoretical reassessment. *Amer. Nat.* 147: 466–491.

Gavrilets, S., H. Li, and M. D. Vose. 1998. Rapid parapatric speciation on holey adaptive landscapes. *Proc. R. Soc. B* 265: 1–7.

Gavrilets, S., and M. D. Vose. 2005. Dynamic patterns of adaptive radiations. *Proc. Natl. Acad. Sci. USA* 102: 18040–18045.

Geyer, L. B., and S. R. Palumbi. 2003. Reproductive character displacement and the genetics of gamete recognition in tropical sea urchins. *Evolution* 57: 1049–1060.

Gibbs, H. L. 1990. Cultural evolution of male song types in Darwin's medium ground finches, *Geospiza fortis*. *Anim. Behav.* 39: 253–263.

Gibbs, H. L., and P. R. Grant. 1987a. Ecological consequences of an exceptionally strong El Niño event on Darwin's finches. *Ecology* 68: 1735–1741.

Gibbs, H. L., and P. R. Grant. 1987b. Oscillating selection on Darwin's finches. *Nature* 327: 511–513.

Gill, F. B. 2004. Blue-winged Warblers (*Vermivora pinus*) versus Golden-winged Warblers (*V. chrysoptera*). *Auk* 121: 1014–1018.

Gill, F. B., and B. G. Murray Jr. 1972. Discrimination behavior and hybridization of the Blue-winged and Golden-winged Warblers. *Evolution* 26: 282–293.

Gillespie, R. G. 2004. Community assembly through adaptive radiation in Hawaiian spiders. *Science* 303: 356–359.

Gingerich, P. D. 2003. Land-to-sea transition of early whales: evolution of Eocene Archaeoceti (Cetacea) in relation to skeletal proportions and locomotion of living semiaquatic mammals. *Paleobiology* 29: 429–454.

Gittenberger, E. 1991. What about non-adaptive radiation? *Biol. J. Linn. Soc.* 43: 263–272.

Givnish, T. J., and K. J. Sytsma, eds. 1997. *Molecular evolution and adaptive radiation.* Cambridge University Press, Cambridge, U.K.

Glynn, P. W., ed. 1990. *Global ecological consequences of the 1982–83 El Niño-Southern Oscillation.* Elsevier, Amsterdam, Netherlands.

Gompert, Z., J. A. Fordyce, M. L. Forister, A. M. Sharpio, and C. C. Nice. 2006. Homoploid hybrid speciation in an extreme habitat. *Science* 314: 1923–1925.

Goodman, D. 1972. *The paleoecology of the Tower Island bird colony: a critical examination of the stability-complexity theory.* Unpubl. Ph.D. thesis, Ohio State University, Columbus, OH.

Gould, S. J. 2002. *The structure of evolutionary theory.* Harvard University Press, Cambridge, MA.

Grant, B. R., and P. R. Grant, 1979. Darwin's finches: population variation and sympatric speciation. *Proc. Natl. Acad. Sci. USA* 76: 2359–2363.

Grant, B. R., and P. R. Grant. 1982. Niche shifts and competition in Darwin's finches: *Geospiza conirostris* and congeners. *Evolution* 36: 637–657.

Grant, B. R., and P. R. Grant, 1989. *Evolutionary dynamics of a natural population: the large cactus finch of the Galápagos.* University of Chicago Press, Chicago, IL.

Grant, B. R., and P. R. Grant, 1993. Evolution of Darwin's finches caused by a rare climatic event. *Proc. R. Soc. B* 251: 111–117.

Grant, B. R., and P. R. Grant, 1996b. Cultural inheritance of song and its role in the evolution of Darwin's finches. *Evolution* 50: 2471–2487.

Grant, B. R., and P. R. Grant, 1996c. High survival of Darwin's finch hybrids: effects of beak morphology and diets. *Ecology* 77: 500–509.

Grant, B. R., and P. R. Grant, 1998a. Hybridization and speciation in Darwin's finches: the role of sexual imprinting on a culturally transmitted trait. In D. J. Howard and S. H. Berlocher, eds., *Endless forms: Species and speciation*, 404–422. Oxford University Press, New York, NY.

Grant, B. R., and P. R. Grant, 2002b. Simulating secondary contact in allopatric speciation: an empirical test of premating isolation. *Biol. J. Linn. Soc.* 76: 545–556.

Grant, B. R., and P. R. Grant, 2002d. Lack of premating isolation at the base of a phylogenetic tree. *Amer. Nat.* 160: 1–19.

Grant, B. R., and P. R. Grant, 2003. What Darwin's finches can teach us about the evolutionary origins and regulation of biodiversity. *BioScience* 53: 965–975.

Grant, P. R. 1972. Convergent and divergent character displacement. *Biol. J. Linn. Soc.* 4: 39–68.

Grant, P. R. 1981a. Speciation and the adaptive radiation of Darwin's finches. *Amer. Sci.* 69: 653–663.

Grant, P. R. 1981b. The feeding of Darwin's finches on *Tribulus cistoides* (L.) seeds. *Anim. Behav.* 29: 785–793.

Grant, P. R. 1993. Hybridization of Darwin's finches on Isla Daphne Major, Galápagos. *Philos. Trans. R. Soc. B* 340: 127–139.

Grant, P. R. 1994. Population variation and hybridization: Comparison of finches from two archipelagos. *Evol. Ecol.* 8: 598–617.

Grant, P. R. 1999. *Ecology and evolution of Darwin's finches*, 2nd ed. Princeton University Press, Princeton, NJ.

Grant, P. R. 2000a. R.C.L. Perkins and evolutionary radiations on islands. *Oikos* 89: 195–201.

Grant, P. R. 2000b. What does it mean to be a naturalist at the end of the twentieth century? *Amer. Nat.* 155: 1–12.

Grant, P. R. 2001. Reconstructing the evolution of birds on islands: 100 years of research. *Oikos* 92: 385–403.

Grant, P. R. 2002. Founder effects and silvereyes. *Proc. Natl. Acad. Sci. USA* 99: 7818–7820.

Grant, P. R., I. Abbott, D. Schluter. R. L. Curry, and L. K. Abbott. 1985. Variation in the size and shape of Darwin's finches. *Biol. J. Linn. Soc.* 25: 1–39.

Grant, P. R., and B. R. Grant. 1980. The breeding and feeding characteristics of Darwin's Finches on Isla Genovesa, Galápagos. *Ecol. Monogr.* 50: 381–410.

Grant, P. R., and B. R. Grant. 1992. Hybridization of bird species. *Science* 256: 193–197.

Grant, P. R., and B. R. Grant. 1994. Phenotypic and genetic effects of hybridization in Darwin's finches. *Evolution* 48: 297–316.

Grant, P. R., and B. R. Grant. 1995. The founding of a new population of Darwin's finches. *Evolution* 49: 229–240.

Grant, P. R., and B. R. Grant. 1996a. Speciation and hybridization of island birds. *Philos. Trans. R. Soc. B* 351: 765–772.

Grant, P. R., and B. R. Grant. 1997a. Hybridization, sexual imprinting and mate choice. *Amer. Nat.* 149: 1–28.

Grant, P. R., and B. R. Grant. 1997b. Mating patterns of Darwin's finch hybrids determined by song and morphology. *Biol. J. Linn. Soc.* 60: 317–343.

Grant, P. R., and B. R. Grant. 1997c. Genetics and the origin of bird species. *Proc. Natl. Acad. Sci. USA* 94: 7768–7775.

Grant, P. R., and B. R. Grant. 1997d. The rarest of Darwin's finches. *Conserv. Biol.* 11: 119–126.

Grant, P. R., and B. R. Grant. 1998b. Speciation and hybridization in island birds. In P. R. Grant, ed., *Evolution on islands*, 142–162. Oxford University Press, Oxford, U.K.

Grant, P. R., and B. R. Grant. 2000. Quantitative genetic variation in populations of Darwin's finches. In T. A. Mousseau, B. Sinervo, and J. Endler, eds., *Adaptive variation in the wild*, 3–40. Academic Press, New York, NY.

Grant, P. R., and B. R. Grant. 2002a. Adaptive radiation of Darwin's finches. *Amer. Sci.* 90: 130–139.

Grant, P. R., and B. R. Grant. 2002c. Unpredictable evolution in a 30-year study of Darwin's finches. *Science* 296: 707–711.

Grant, P. R., and B. R. Grant. 2006a. Evolution of character displacement in Darwin's finches. *Science* 313: 224–226.

Grant, P. R., and B. R. Grant. 2006b. Species before speciation is complete. *Ann. Missouri Bot. Gard.* 93: 94–102.

Grant, P. R., B. R. Grant, L. F. Keller, J. A. Markert, and K. Petren. 2003. Inbreeding and interbreeding in Darwin's finches. *Evolution* 57: 2911–2916.

Grant, P. R., B. R. Grant, L. F. Keller, and K. Petren. 2005b. Extinction behind our backs: the possible fate of one of the Darwin's finch species on Isla Floreana, Galápagos. *Biol. Conserv.* 122: 499–503.

Grant, P. R., B. R. Grant, J. A. Markert, L. F. Keller, and K. Petren. 2004. Convergent evolution of Darwin's finches caused by introgressive hybridization and selection. *Evolution* 58: 1588–1599.

Grant, P. R., B. R. Grant, and K. Petren. 2000. The allopatric phase of speciation: the sharp-beaked ground finch (*Geospiza difficilis*) on the Galápagos islands. *Biol. J. Linn. Soc* 69: 287–317.

Grant, P. R., B. R. Grant, and K. Petren. 2001. A population founded by a single pair of individuals: establishment, expansion, and evolution. *Genetica* 112/113: 359–382.

Grant, P. R., B. R. Grant, and K. Petren. 2005a. Hybridization in the recent past. *Amer. Nat.* 166: 56–67.

Grant, P. R., and T. D. Price. 1981. Population variation in continuously varying traits as an ecological genetics problem. *Syst. Zool.* 21: 795–811.

Gray, A. P. 1958. *Bird hybrids.* Commonwealth Agricultural Bureaux, Farnham Royal, UK.

Grinnell, J. 1924. Geography and evolution. *Ecology* 5: 225–229.

Gulledge, J. L. 1970. *An analysis of song in the mockingbird genera Nesomimus and Mimus.* Unpubl. M.A. thesis, San Francisco State University, San Francisco, CA.

Hajibabaei, M., D. H. Janzen, J. M. Burns, W. Hallwechs, and P.D.N. Hebert. 2006. DNA barcodes distinguish species of tropical Lepidoptera. *Proc. Natl. Acad. Sci. USA* 103: 968–971.

Haldane, J.B.S. 1922. Sex ratio and unisexual sterility in animal hybrids. *J. Genet.* 12: 101–109.

Hall, B. P., and R. E. Moreau. 1970. *An atlas of speciation in African passerine birds.* British Museum of Natural History, London, U.K.

Hamann, O. 1981. Plant communities of the Galápagos Islands. *Dansk Botanisk Archiv.* 34: 1–163.

Harmon, L. J., J. J. Kolbe, J. M. Cheverud, and J. B. Losos. 2005. Convergence and the multidimensional niche. *Evolution* 59: 409–421.

Harmon, L. J., J. A. Schutte II, A. Larson, and J. B. Losos. 2003. Tempo and mode of evolutionary radiation in iguanian lizards. *Science* 301: 961–964.

Harrison, R. G. 1998. Linking pattern and process: the relevance of species concepts for the study of speciation. In D. J. Howard and S. H. Berlocher, eds., *Endless forms: Species and speciation,* 19–31. Oxford University Press, New York, NY.

Harrison, R. G., ed. 1993. *Hybrid zones and the evolutionary process.* Oxford University Press, New York, NY.

Haug, G. H., D. M. Sigman, R. Tiedemann, T. F. Pedersen, and M. Sarnthein. 1999. Onset of permanent stratification in the subarctic Pacific Ocean. *Nature* 401: 779–782.

Hebets, E. A. 2003. Subadult experience influences adult mate choice in an arthropod: exposed female wolf spiders prefer males of a familiar phenotype. *Proc. Natl. Acad. Sci. USA* 100: 13390–13395.

Hedges, S. B. 1989. Evolution and biogeography of West Indian frogs of the genus *Eleutherodactylus*: slow-evolving loci and the major groups. In C. A. Woods, ed., *Biogeography of the West Indies: Past present and future,* 305–370. Sandhill Crane Press, Gainesville, FL.

Hedges, S. B., C. A. Hess, and L. R. Maxon. 1992. Caribbean biogeography: molecular evidence for dispersal in West Indian terrestrial vertebrates. *Proc. Natl. Acad. Sci. USA* 89: 1909–1913.

Hedrick, P.W. 1998. *Genetics of populations,* 2nd ed. Jones and Bartlett, Sudbury, MA.

Helbig, A. J., A. G. Knox, D. T. Parkin, G. Sangster, and M. Collinson. 2002. Guidelines for assigning species rank. *Ibis* 144: 518–525.

Hendry, A. P., P. R. Grant, B. R. Grant, H. A. Ford, M. J. Brewer, and J. Podos. 2006. Possible human impacts on adaptive radiation in Darwin's finches. *Proc. R. Soc. B* 273: 1187–1194.

Herder, F., A. W. Nolte, J. Pfaender, J. Schwarzer, R. K. Hadiaty, and U. K. Schliewen. 2006. Adaptive radiation and hybridization in Wallace's Dreamponds: evidence from sailfin silversides in the Malili Lakes of Sulawesi. *Proc. R. Soc. B* 273: 2209–2217.

Herrel, A., J. Podos, S. K. Huber, and A. P. Hendry. 2005a. Bite performance and morphology in a population of Darwin's finches: Implications for the evolution of beak shape. *Funct. Ecol.* 19: 43–48.

Herrel, A., J. Podos, S. K. Huber, and A. P. Hendry. 2005b. Evolution of bite force in Darwin's finches: A key role for head width. *J. Evol. Biol.* 18: 669–675.

Higashi, M., G. Takimoto, and N. Yamamura. 1999. Sympatric speciation by sexual selection. *Nature* 402: 523–526.

Ho, S. Y., M. J. Phillips, A. Cooper, and A. J. Drummond. 2005. Time dependency of molecular rate estimates and systematic overestimation of recent divergence times. *Mol. Biol. Evol.* 22: 1561–1568.

Horner-Devine, M. C., K. M. Carney, and B.J.M. Bohannan. 2004. An ecological perspective on bacterial diversity. *Proc. R. Soc. B* 271: 113–122.

Hoskin, C. J., M. Higgie, K. R. McDonald, and C. Moritz. 2005. Reinforcement drives rapid allopatric speciation. *Nature* 437: 1353–1356.

Hostert, E. E. 1997. Reinforcement: a new perspective on an old controversy. *Evolution* 51: 697–702.

Howard, R. D. 1974. The influence of sexual selection and interspecific competition on mockingbird song. *Evolution* 28: 428–438.

Howarth, D. G, and D. A. Baum. 2005. Genealogical evidence of homoploid hybrid speciation in an adaptive radiation of *Scaevola* (Goodeniaceae) in the Hawaiian islands. *Evolution* 59: 948–961.

Huber, S. K., and J. Podos. 2006. Beak morphology and song features covary in a population of Darwin's finches (*Geospiza fortis*). *Biol. J. Linn. Soc.* 88: 489–498.

Hudson, R. R., and J. A. Coyne. 2002. Mathematical consequences of the genealogical concept. *Evolution* 56: 1557–1565.

Hundley, M. H. 1963. Notes on methods of feeding and the use of tools in the Geospizinae. *Auk* 80: 372–373.

Hunt, J. S., E. Bermingham, and R. E. Ricklefs. 2001. Molecular systematics and biogeography of Antillean thrashers, tremblers, and mockingbirds (Aves: Mimidae). *Auk* 118: 35–55.

Hutchinson, G. E. 1965. *The ecological theater and the evolutionary play*. Yale University Press, New Haven, CT.

Huxley, J. S. 1938. Species formation and geographical isolation. *Proc. Linn. Soc. Lond.* 150: 253–264.

Huxley, J. S. 1942. *Evolution, the modern synthesis*. Allen & Unwin, London.,U.K.

187

Huxley, J. S. ed. 1940. *The new systematics.* Clarendon Press, Oxford, U.K.

Immelmann, K. 1975. Ecological significance of imprinting and early learning. *Annu. Rev. Ecol. Syst.* 6: 15–37.

Irwin, D. E., and T. D. Price. 1999. Sexual imprinting, learning and speciation. *Heredity* 82: 347–354.

Jablonski, D., K. Roy, and J. W. Valentine. 2006. Out of the tropics: evolutionary dynamics of the latitudinal diversity gradient. *Science* 314: 102–106.

Jackson, J.B.C. 1994. Constancy and change of life in the sea. *Philos. Trans. R. Soc. B* 344: 55–60.

James, H. F. 2004. The osteology and phylogeny of the Hawaiian finch radiation (Fringillidae: Drepanidini), including extinct taxa. *Zool. J. Linn. Soc.* 141: 207–255.

Jiggins, C. D., R. Mallarino, K. R. Willmott, and E. Bermingham. 2006. The phylogenetic pattern of speciation and wing pattern change in neotropical. *Ithomia* butterflies (Lepidoptera: Nymphalidae). *Evolution* 60: 1454–1466.

Johnson, M. P., and P. H. Raven. 1973. Species number and endemism: the Galápagos archipelago. *Science* 179: 893–895.

Johnson, M. S., J. Murray, and B. Clarke. 2000. Parallel evolution in Marquesan partulid land snails. *Biol. J. Linn. Soc.* 69: 577–598.

Joyce, D. A., D. H. Lunt, R. Bills, G. F. Turner, C. Katongo, N. Duftner, C. Sturmbauer, and O. Seehausen. 2005. An extant cichlid fish radiation emerged in an extinct Pleistocene lake. *Nature* 435: 90–95.

Kambysellis, M. P., and E. M. Craddock. 1997. Ecological and reproductive shifts in the diversification of the endemic Hawaiian *Drosophila.* In T. J. Givnish and K. J. Sytsma, eds., *Molecular evolution and adaptive radiation,* 475–509. Cambridge University Press, Cambridge, U.K.

Kaneshiro, K. Y., R. G. Gillespie, and H. L. Carson. 1995. Chromosomes and male genitalia of Hawaiian *Drosophila*: tools for interpreting phylogeny and geography. In W. W. Wagner and V. A. Funk, eds., Hawaiian biogeography: *evolution on a hotspot,* 57–71. Smithsonian Institution Press, Washington, DC.

Kawecki, T. 1997. Sympatric speciation by habitat specialization driven by deleterious mutations. *Evolution* 51: 1751–1763.

Keller, L. F., P. R. Grant, B. R. Grant, and K. Petren. 2001. Heritability of morphological traits in Darwin's finches: misidentified paternity and maternal effects. *Heredity* 87: 325–336.

Keller, L. F., P. R. Grant, B. R. Grant, and K. Petren. 2002. Environmental conditions affect the magnitude of inbreeding depression in survival of Darwin's finches. *Evolution* 56: 1229–1239.

Kerr, R. A. 2001. The tropics return to the climate system. *Science* 292: 660–661.

Kirchman, J. J., S. J. Hackett, S. M. Goodman, and J. M. Bates. 2001. Phylogeny and systematics of ground rollers (Brachypteraciidae) of Madagascar. *Auk* 118: 849–863.

Kirzian, D., A. Trager, M. A. Donnelly, and J. W. Wright. 2004. Evolution of Galápagos island lava lizards (Iguania: Tropiduridae: *Microlophus*). *Mol. Phylogenet. Evol.* 32: 761–769.

Kleindorfer, S., T. W. Chapman, H. Winkler, and F. J. Sulloway. 2006. Adaptive divergence in contiguous populations of Darwin's small ground finch (*Geospiza fuliginosa*). *Evol. Ecol. Res.* 8: 357–372.

Kocher, T. D. 2004. Adaptive evolution and explosive speciation: the cichlid fish model. *Nature Revs. Genet.* 5: 288–298.

Kondrashov, A. S., and F. A. Kondrashov. 1999. Interactions among quantitative traits in the course of sympatric speciation. *Nature* 400: 351–354.

Kozak, K. H., D. W. Weisrock, and A. Larson. 2006. Rapid lineage accumulation in a non-adaptive radiation: phylogenetic analysis of diversification rates in eastern North American woodland salamanders (Plethodontidae: *Plethodon*). *Proc. R. Soc. B* 273: 539–546.

Labandeira, C. C., D. L. Dilcher, D. R. Davis, and D. L. Wagner. 1994. Ninety-seven million years of angiosperm-insect association: paleobiological insights into the meaning of coevolution. *Proc. Natl. Acad. Sci. USA* 91: 12278–12282.

Lack, D. 1945. The Galápagos finches (Geospizinae): a study in variation. *Occas. Pap. Calif. Acad. Sci.* 21: 1–159.

Lack, D. 1947. *Darwin's finches.* Cambridge University Press, Cambridge, U.K.

Lambeck, K. and J. Chappell. 2001. Sea level change through the last glacial cycle. *Science* 292: 679–686.

Laurie, C. C. 1997. The weaker sex is heterogametic: 75 years of Haldane's rule. *Genetics* 147: 937–951.

Lawrence, K. T., L. Zhonghui, and T. D. Herbert. 2006. Evolution of the Eastern Tropical Pacific through Plio-Pleistocene glaciation. *Science* 312: 79–83.

Lea, D. W., D. K. Pak, C. L. Belanger, H. J. Spero, M. A. Hall, and N. J. Shackleton. 2006. Paleoclimate history of Galápagos surface waters over the last 135,000 yr. *Quaternary Sci. Revs.* 25: 1152–1167.

Lewontin, R. C., and L. C. Birch. 1966. Hybridization as a source of variation for adaptation to new environments. *Evolution* 20: 315–336.

Liou, L. W., and T. D. Price. 1994. Speciation by reinforcement of premating isolation. *Evolution* 48: 1451–1459.

Lopez, T. J., D. E. Hauselman, L. R. Maxon, and J. W. Wright. 1992. Preliminary analysis of phylogenetic relationships among Galápagos Island lizards of the genus *Tropidurus*. *Amphibia-Reptilia* 13: 327–339.

Losos, J. B. 1994. Integrative approaches to evolutionary ecology: *Anolis* lizards as model systems. *Annu. Rev. Ecol. Syst.* 25: 467–493.

Losos, J. B. 1998. Ecological and evolutionary determinants of the species-area relationship in Caribbean anoline lizards. In P. R. Grant, ed., *Evolution on islands*, 210–224. Oxford University Press, Oxford, U.K.

Losos, J. B., T. R. Jackman, A. Larson, K. de Queiroz, and L. Rodriguez-Schettino. 1998. Contingency and determinism in replicated adaptive radiations of island lizards. *Science* 279: 2115–2118.

Lovette I. J. 2004. Mitochondrial dating and mixed support for the "2% rule" in birds. *Auk* 121: 1–6.

Lovette, I. J., and E. Bermingham. 1999. Explosive speciation in the New World *Dendroica* warblers. *Proc. R. Soc. B* 266: 1629–1636.

Lovette, I. J., E. Bermingham, and R. E. Ricklefs. 2002. Clade-specific morphological diversification and adaptive radiation in Hawaiian songbirds. *Proc. R. Soc. B* 269: 37–42.

Lukhtanov, V. A., N. P. Kandul, J. B. Plotkin, A. V. Dantchenko, D. Haig, and N. E. Pierce. 2005. Reinforcement of pre-zygotic isolation and karyotype evolution in *Agrodiaetus* butterflies. *Nature* 436: 385–389.

Lynch, A., and A. J. Baker. 1990. Increased vocal discrimination by learning: sympatry in two species of chaffinches. *Behaviour* 116: 109–126.

MacFadden, B. J., and R. C. Hulbert. 1988. Explosive speciation at the base of the adaptive radiation of Miocene grazing horses. *Nature* 336: 466–468.

Mallett, J. 2005. Hybridization as an invasion of the genome. *Trends Ecol. Evol.* 20: 229–237.

Mallet, J., W. O. McMillan, and C. D. Jiggins. 1998. Mimicry and warning color at the boundary between races and species. In D. J. Howard and S. H. Berlocher, eds., *Endless Forms: species and speciation*, 390–403. Oxford University, New York, NY.

Marshall, D. C., and J. R. Cooley. 2000. Reproductive character displacement and speciation in periodical cicadas, with a description of a new species, 13-year *Magicicada neotredecim*. *Evolution* 54: 1313–1325.

Martinsen, G. D., T. G. Whitham, R. J. Turek, and P. Keim. 2001. Hybrid populations selectively filter gene introgression. *Evolution* 55: 1325–1335.

Matyjasiak, P. 2005. Birds associate species-specific acoustic and visual cues: recognition of heterospecific rivals by male blackcaps. *Behav. Ecol.* 16: 467–471.

Maynard Smith, J. 1966. Sympatric speciation. *Amer. Nat.* 100: 637–650.

Mayr, E. 1942. *Systematics and the origin of species.* Columbia University Press, New York, NY.

Mayr, E. 1954. Change in genetic environment and evolution. In J. Huxley, A. C. Hardy, and E. B. Ford, eds., *Evolution as a process*, 157–180. Allen & Unwin, London, UK.

Mayr, E. 1963. *Animal species and evolution.* Belknap Press, Harvard, Cambridge, MA.

Mayr, E. 1992. Controversies in retrospect. In D. J. Futuyma and J. Antonovics, eds., *Oxford surveys in evolutionary biology*, 1–34. Oxford University Press, Oxford, UK.

Mayr, E. 2004. *What makes biology unique? Considerations on the autonomy of a scientific discipline.* Cambridge University Press, Cambridge, U.K.

Mayr, E., and J. Diamond. 2001. *Birds of Northern Melanesia*. Oxford University Press, New York, NY.

McPhaden, M. J., S. E. Zebiak, and M. H. Glantz. 2006. ENSO as an integrating concept in earth science. *Science* 314: 1740–1745.

Merrell, D. J. 1994. *The adaptive seascape*. University of Minnesota Press, Minneapolis, MN.

Miller, A. J., D. R. Cayan, T. P. Barnett, N. E. Graham, and J. M. Oberhuber. 1994. The 1976–77 climate shift in the Pacific Ocean. *Oceanogr.* 7: 21–26.

Millikan, G. C., and R. I. Bowman. 1967. Observations on Galápagos tool-using finches in captivity. *Living Bird* 6: 23–41.

Muller, H. J. 1940. Bearing of the *Drosophila* work on systematics. In J. Huxley, ed., *The new systematics*, 185–286. Clarendon Press, Oxford, U.K.

Nee, S. 2006. Birth-death models in macroevolution. *Annu. Rev. Ecol. Evol. Syst.* 37: 1–17.

Nee, S., R. M. May, and P. H. Harvey. 1994. The reconstructed evolutionary process. *Philos. Trans. R. Soc. B* 344: 305–311.

Newton, I. 2003. The *speciation and biogeography of birds*. Academic Press, London, U.K.

Nosil, P., B. J. Crespi, and C. P. Sandoval. 2003. Reproductive isolation driven by the combined effects of ecological adaptation and reinforcement. *Proc. R. Soc. B* 270: 1911–1918.

Olson, S. L., and H. F. James. 1981. Fossil birds from the Hawaiian Islands: evidence for wholesale extinction by man before western contact. *Science* 217: 633–635.

Orr, H. A. 1996. Dobzhansky, Bateson, and the genetics of speciation. *Genetics* 144: 1331–1335.

Osborn, H. F. 1900. The geological and faunal relations of Europe and America during the Tertiary Period and the theory of successive invasions of an African fauna. *Science* 11: 563–564.

Panov, E. N. 1989. *Natural hybridisation and ethological isolation in birds*. Nauka, Moscow, Russia.

Parent, C. E., and B. Crespi. 2006. Sequential colonization and diversification of Galápagos endemic land snail genus *Bulimulus* (Gastropoda, Stylommatophora). *Evolution* 60: 2311–2328.

Patten, M. A., J. T. Rotenberry, and M. Zuk. 2004. Habitat selection, acoustic adaptation, and the evolution of reproductive isolation. *Evolution* 58: 2144–2155.

Patterson, D. J. 1999. The diversity of eukaryotes. *Amer. Nat.* 154 supplement: S96–S124.

Patterson, N., D. J. Richter, S. Gnerre, E. S. Lander, and D. Reich. 2006. Genetic evidence for complex speciation of humans and chimpanzees. *Nature* 441: 1103–1108.

Payne, R. B. 1973. Behavior, mimetic songs and song dialects, and relationships of parasitic Indigobirds (*Vidua*) of Africa. *Ornithol. Monogr.* 11: 1–333.

Payne, R. B., L. L. Payne, J. L. Woods, and M. D. Sorenson. 2000. Imprinting and the origin of parasite-host associations in brood-parasitic indigobirds, *Vidua chalybeata*. *Anim. Behav.* 59: 69–81.

Peck, S. B. 1996. Diversity and distribution of orthopteroid insects of the Galápagos Islands, Ecuador. *Canad. J. Zool.* 74: 1497–1510.

Pemberton, R. W., and G. S. Wheeler. 2006. Orchid bees don't need orchids: evidence from the naturalization of an orchid bee in Florida. *Ecology* 87: 1995–2001.

Perkins, R.C.L. 1903. Vertebrata. In D. Sharp, ed., *Fauna Hawaiiensis*, 365–466. Cambridge University Press, Cambridge, U.K.

Perkins, R.C.L. 1913. Introduction. In D. Sharp, ed., *Fauna Hawaiiensis*, xv–ccxxviii. Cambridge University Press, Cambridge, U.K.

Peterson, M. A., B. A. Honchak, S. E. Locke, T. E. Beeman, J. Mendoza, J. Green, K. J. Buckingham, M. A. White, and K. J. Monsen. 2005. Relative abundance and the species-specific reinforcement of male mating preference in the *Chrysochus* (Coleoptera: Chysomelidae) hybrid zone. *Evolution* 59: 2639–2655.

Petren, K., B. R. Grant, and P. R. Grant. 1999. A phylogeny of Darwin's finches based on microsatellite DNA length variation. *Proc. R. Soc. B* 266: 321–329.

Petren, K., P. R. Grant, B. R. Grant, and L. F. Keller. 2005. Comparative landscape genetics and the adaptive radiation of Darwin's finches: the role of peripheral isolation. *Mol. Ecol.* 14: 2943–2957.

Pfennig, K. 2003. A test of alternative hypotheses for the evolution of reproductive isolation between spadefoot toads: support for the reinforcement hypothesis. *Evolution* 57: 2842–2851.

Podos, J. 2001. Correlated evolution of morphology and vocal signal structure in Darwin's finches. *Nature* 409: 185–188.

Podos, J., S. K. Huber, and B. Taft. 2004a. Bird song: the interface of evolution and mechanism. *Annu. Rev. Ecol. Syst.* 35: 55–87.

Podos, J., and S. Nowicki. 2004. Beaks, adaptation, and vocal evolution in Darwin's Finches. *BioScience* 54: 501–510.

Podos, J., J. A. Southall, and M. R. Rossi-Santos. 2004b. Vocal mechanics in Darwin's finches: correlation of beak gape and song frequency. *J. Exp. Biol.* 207: 607–619.

Porter, D. M. 1976. Geography and dispersal of Galapagos Islands vascular plants. *Nature* 264: 745–746.

Prager, E. M., and A. C. Wilson. 1975. Slow evolutionary loss of the potential for interspecific hybridization in birds: a manifestation of slow regulatory evolution. *Proc. Natl. Acad. Sci. USA* 72: 200–204.

Pratt, H. D. 2005. *The Hawaiian honeycreepers* Drepanidinae. Oxford University Press, Oxford, U.K.

Price, T. D. 1987. Diet variation in a population of Darwin's finches. *Ecology* 68: 1015–1028.

Price, T. D. 1998. Sexual selection and natural selection in bird speciation. *Philos. Trans. R. Soc. B* 353: 251–260.

Price, T. D. 2007. *Speciation in birds.* Roberts & Co., Greenwood Village, CO.

Price, T. D., and M. M. Bouvier. 2002. The evolution of F_1 postzygotic incompatibilities in birds. *Evolution* 56: 2083–2089.

Price, T. D., H. L. Gibbs, L. de Sousa, and A. D. Richman. 1998. Different timings of the adaptive radiations of North American and Asian warblers. *Proc. R. Soc. B* 265: 1969–1975.

Price, T. D., P. R. Grant, H. L. Gibbs, and P. T. Boag. 1984. Recurrent patterns of natural selection in a population of Darwin's finches. *Nature* 309: 787–789.

Price, T. D., I. Lovette, E. Bermingham, H. L. Gibbs, and A. D. Richman. 2000. The imprint of history on communities of North American and Asian warblers. *Amer. Nat.* 156: 354–367.

Price, T. D., A. Qvarnström, and D. E. Irwin. 2003. The role of phenotypic plasticity in driving genetic evolution. *Proc. R. Soc. B* 270: 1433–1440.

Pritchard, J. K., M. Stephens, and P. Donnelly. 2000. Inference of population structure using multilocus genotype data. *Genetics* 155: 945–959.

Prodon, R., J.-C. Thibault, and P.-A. Dejaifve. 2002. Expansion vs compression of bird altitude ranges on a Mediterranean island. *Ecology* 83: 1294–1306.

Provine, W. B. 1989. Founder effects and genetic revolutions in microevolution and speciation: a historical perspective. In L. V. Giddings, K. Y. Kaneshiro, and W. W. Anderson, eds., *Genetics, speciation, and the founder principle*, 43–76. Oxford University Press, New York, NY.

Rabosky, D. L. 2006. Likelihood methods for detecting temporal shifts in diversification rates. *Evolution* 60: 1152–1164.

Rasmann, C. 1997. Evolutionary age of the Galápagos iguanas predates the age of the present Galápagos islands. *Mol. Phylogenet. Evol.* 7: 158–172.

Ratcliffe, L. M. 1981. *Species recognition in Darwin's ground finches (Geospiza, Gould).* Unpubl. Ph. D. thesis, McGill University, Montreal, Canada.

Ratcliffe, L. M., and P. R. Grant. 1983a. Species recognition in Darwin's finches (*Geospiza*, Gould). I. Discrimination by morphological cues. *Anim. Behav.* 31: 1139–1153.

Ratcliffe, L. M., and P. R. Grant. 1983b. Species recognition in Darwin's finches (*Geospiza*, Gould). II. Geographic variation in mate preference. *Anim. Behav.* 31: 1154–1165.

Ratcliffe, L. M., and P. R. Grant. 1985. Species recognition in Darwin's Finches (*Geospiza*, Gould). III. Male responses to playback of different song types, dialects and heterospecific songs. *Anim. Behav.* 33: 290–307.

Raymo, M. E., K. Ganley, S. Carter, D. W. Oppo, and J. McManus. 1998. Millenial-scale climate instability during the early Pleistocene epoch. *Nature* 392: 699–703.

Rensch, B. 1933. Zoologische Systematik und Artbildungsproblem. *Zool. Anzeiger,* Suppl. 6: 19–83.

Reudink, M. W., S. G. Mech, and R. L. Curry. 2005. Extrapair paternity and mate choice in a chickadee hybrid zone. *Behav. Ecol.* 17: 56–62.

Rice, W. R., and E. E. Hostert. 1993. Perspective: Laboratory experiments on specia-tion: what have we learned in forty years? *Evolution* 47: 1637–1653.

Richman, A. D. 1996. Ecological diversification and community structure in the Old World leaf warblers (Genus *Phylloscopus*): a phylogenetic perspective. *Evolution* 50: 2461–2470.

Richman, A. D., and T. D. Price. 1992. Evolution of ecological differences in the Old World leaf warblers. *Nature* 355: 817–821.

Ricklefs, R. E., and G. W. Cox. 1972. Taxon cycles in the West Indian avifauna. *Amer. Nat.* 106: 195–219.

Ricklefs, R. E., and D. Schluter. 1993. Species diversity: regional and historical influ-ences. In R. E. Ricklefs and D. Schluter, eds., *Species diversity in ecological commu-nities: historical and geographical perspectives,* 350–363. University of Chicago Press, Chicago, IL.

Ridgway, R. S. 1901. *The birds of North and Middle America,* vol. 1. Govt. Printing Office, Washington, DC.

Riebel, K. 2000. Early exposure leads to repeatable preferences for male song in female zebra finches. *Proc. R. Soc. B* 267: 2553–2558.

Riebel, K., I. M. Smallegange, N. J. Terpstra, and J. J. Bolhuis. 2002. Sexual equality in zebra finch song preference: evidence for a dissociation between song recognition and production learning. *Proc. R. Soc. B* 269: 729–733.

Rieseberg, L. H. 1997. Hybird orgins of plant species. *Annu. Rev. Ecol. Syst.* 28: 359–389.

Rieseberg, L. H., O. Raymond, D. M. Rosenthal, Z. Lai, K. Livingstone, T. Nakazato, J. L. Durphy, A. E. Schwartzbach, L. A. Donovan, and C. Lexer. 2003. Major ecologi-cal transitions in wild sunflowers facilitated by hybridization. *Science* 301: 1211–1216.

Rowher, S., E. Bermingham, and C. Wood. 2001. Plumage and mitochondrial DNA haplotype variation across a moving hybrid zone. *Evolution* 55: 405–422.

Roy, M. S. 1997. Recent diversification in African greenbuls (Pycnonotidae: *Andropadus*) supports a montane speciation model. *Proc. R. Soc. B* 264: 1337–1344.

Ruta, M., P. J. Wagner, and M. J. Coates. 2006. Evolutionary patterns in early tetrapods. I. Rapid initial diversification followed by decrease in rates of character change. *Proc. R. Soc. B* 273: 2107–2111.

Ryan, P. G., C. L. Moloney, and J. Hudon. 1994. Color variation and hybridization among *Nesospiza* buntings on Inaccessible island, Tristan de Cunha. *Auk* 111: 314–327.

Sætre, G.-P., T. Borge, J. Lindell, T. Moum, C. R. Primmer, B. C. Sheldon, J. Haavie, A. Johnson, and H. Ellegren. 2001. Speciation, introgressive hybridization and non-linear rate of molecular evolution in flycatchers. *Mol. Ecol.* 10: 737–749.

Sætre, G.-P., T. Borge, K. Lindroos, J. Haavie, B. C. Sheldon, C. R. Primmer, and A. C. Syvänen. 2003. Sex chromosome evolution and speciation in Ficedula flycatchers. *Proc. R. Soc. B* 270: 53–59.

Sætre, G.-P., T. Moum, S. Bureš, M. Král, M. Adamjan, and J. Moreno. 1997. A sexually selected character displacement in flycatchers reinforces premating isolation. *Nature* 387: 589–592.

Sato, A., C. O'hUigin, F. Figueroa, P. R. Grant, B. R. Grant, and J. Klein. 1999. Phylogeny of Darwin's finches as revealed by mtDNA sequences. *Proc. Natl. Acad. Sci. USA* 96: 5101–5106.

Sato, A., H. Tichy, C. O'hUigin, P. R. Grant, B. R. Grant, and J. Klein. 2001. On the origin of Darwin's finches. *Mol. Biol. Evol.* 18: 299–311.

Schliewen, U. K., D. Tautz, and S. Pääbo. 1994. Sympatric speciation suggested by monophyly of crater lake cichlids. *Nature* 368: 629–632.

Schluter, D. 1996. Ecological causes of adaptive radiation. *Amer. Nat.* 148 (suppl.): S40–S64.

Schluter, D. 1998. Ecological causes of speciation. In D. J. Howard and S. H. Berlocher, eds., Endless forms: species and speciation, 114–129. Oxford University Press, New York, NY.

Schluter, D. 2000. *The ecology of adaptive radiation.* Oxford University Press, Oxford, U.K.

Schluter, D., and P. R. Grant. 1984a. Determinants of morphological patterns in communities of Darwin's finches. *Amer. Nat.* 123: 175–196.

Schluter, D., and P. R. Grant. 1984b. Ecological correlates of morphological evolution in a Darwin's Finch species. *Evolution* 38: 856–869.

Schluter, D., T. D. Price, and P. R. Grant. 1985. Ecological character displacement in Darwin's finches. *Science* 277: 1056–1059.

Schwarz, D., B. M. Matta, N. L. Shakir-Botteri, and B. A. McPheron. 2005. Host shift to an invasive plant triggers rapid animal hybrid speciation. *Nature* 436: 546–549.

Secondi, J., V. Bretagnolle, C. Compagnon, and B. Faivre. 2003. Species-specific song convergence in a moving hybrid zone between two passerines. *Biol. J. Linn. Soc.* 80: 507–517.

Seddon, N. 2005. Ecological adaptation and species recognition drives vocal evolution in neotropical suboscine birds. *Evolution* 59: 200–215.

Seehausen, O. 2004. Hybridization and adaptive radiation. *Trends Ecol. Evol.* 19: 198–207.

Seehausen, O. 2006. Review. African cichlid fish: a model system in adaptive radiation research. *Proc. R. Soc. B* 273: 1987–1998.

Seehausen, O., J.J.M. van Alphen, and F. Witte. 1997. Cichlid fish diversity threatened by eutrophication that curbs sexual selection. *Science* 277: 1808–1811.

Sequeira, A. S., A. A. Lanteri, M. A. Scataglini, V. A. Confalonieri, and B. D. Farrell. 2000. Are flightless *Galapaganus* weevils older than the Galápagos Islands they inhabit? *Heredity* 85: 20–29.

Severinghaus, L. L., and Y.-M. Kuo. 1994. Mate choice as a cause for unequal hybridization between Chinese and Styan's Bulbuls. *J. für Ornithol.* 135: 363.

Simpson, G. G. 1944. *Tempo and mode of evolution.* Columbia University Press, New York, NY.

Simpson, G. G. 1949. *The meaning of evolution: a study of the history of life and its significance for man.* Yale University Press, New Haven, CT.

Simpson, G. G. 1953. *The major features of evolution.* Columbia University Press, New York, NY.

Sims, R. W. 1959. The *Ceyx erithacus* and *rufidorsus* species problem. *J. Linn. Soc. Lond. Zool.* 44: 212–221.

Sinton, C. W., D. M. Christie, and R. A. Duncan. 1996. Geochronology of Galápagos seamounts. *J. Geophys. Res.* 101: 13689–13700.

Slabbekoorn, H., and T. B. Smith. 2000. Does bill size polymorphism affect courtship characteristics in the African finch *Pyrenestes ostrinus*? *Biol. J. Linn. Soc.* 71: 737–753.

Slagsvold, T., B. T. Hansen, L. E. Johannessen, and J. T. Lifjeld. 2002. Mate choice and imprinting in birds studied by cross-fostering in the wild. *Proc. R. Soc. B* 269: 1449–1456.

Slatkin, M. 1975. Gene flow and the geographic structure of natural populations. *Science* 236: 787–792.

Smith, J.N.M., P. R. Grant, B. R. Grant, I. Abbott, and L. K. Abbott. 1978. Seasonal variation in feeding habits of Darwin's ground finches. *Ecology* 59: 1137–1150.

Sol, D., R. P. Duncan, T. M. Blackburn, P. Cassey, and L. Lefebvre. 2005. Big brains, enhanced cognition, and response of birds to novel environments. *Proc. Natl. Acad. Sci. USA* 102: 5460–5465.

Soltis, D. E., P. S. Soltis, P. K. Endress, and M. W. Chase. 2005. *Phylogeny and evolution of angiosperms.* Sinauer, Sunderland, MA.

Sorenson, M. D., K. M. Sefc, and R. B. Payne. 2003. Speciation by host switch in brood parasitic indigobirds. *Nature* 424: 928–931.

Steadman, D. W. 1986. *Holocene vertebrate fossils from Isla Floreana, Galápagos.* Smithsonian Contributions to Zoology, no. 413.

Stebbins, G. L., Jr. 1959. The role of hybridization in evolution. *Proc. Amer. Philos. Soc.* 103: 231–251.

Stern, D. L., and P. R. Grant. 1996. A phylogenetic reanalysis of allozyme variation among populations of Galápagos finches. *Zool. J. Linn. Soc.* 118: 119–134.

Stresemann, E. 1936. Zur Frage der Artbildung in der Gattung *Geospiza*. *Org. Club Nederl. Vogelkunde* 9: 13–21.

Swarth, H. S. 1931. The avifauna of the Galapagos islands. *Occas. Pap. Calif. Acad. Sci.* 18: 1–299.

Swarth, H. S. 1934. The bird fauna of the Galápagos Islands in relation to species formation. *Biol. Revs.* 9: 213–234.

Tao, Y., S. Chen, D. L. Hartl, and C. C. Laurie. 2003. Genetic dissection of hybrid incompatibilities between *Drosophila simulans* and *D. mauritiana*. I. Differential accumulation of hybrid male sterility effects on the X and autosomes. *Genetics* 164: 1383–1398.

Tarr, C. L., S. Conant, and R. C. Fleischer. 1998. Founder events and variation at microsatellite loci in an insular passerine bird, the Laysan finch (*Telespiza cantans*). *Mol. Ecol.* 7: 719–731.

Taylor, E. B., J. W. Boughman, M. Groenenboom, M. Sniatynski, D. Schluter, and J. Gow. 2006. Speciation in reverse: morphological and genetic evidence of the collapse of a three-spined stickleback (*Gasterosteus aculeatus*) species pair. *Mol. Ecol.* 15: 343–355.

Tebbich, S., and R. Bshary. 2004. Cognitive abilities related to tool use in the woodpecker finch, *Cactospiza pallida*. *Anim. Behav.* 67: 689–697.

Tebbich, S., M. Taborsky, B. Fessl, and D. Blomqvist. 2001. Do woodpecker finches acquire tool use by social learning? *Proc. R. Soc. B* 268: 2189–2193.

Tebbich, S., M. Taborsky, B. Fessl, M. Dvorak, and H. Winkler. 2004. Feeding behavior of four arboreal Darwin's finches: adaptations to spatial and seasonal variability. *Condor* 106: 95–105.

Tegelström, H., and H. P. Gelter. 1990. Haldane's rule and sex-biased gene flow between two hybridizing flycatcher species (*Ficedula albicollis* and *F. hypoleuca*, Aves: Muscicapidae). *Evolution* 44: 2012–2021.

Templeton, A. R. 1989. The meaning of species and speciation: a genetic perspective. In D. Otte and J. A. Endler, eds., *Speciation and its consequences*, 3–27. Sinauer, Sunderland, MA.

ten Cate, C., M. N. Verzijden, and E. Etman. 2006. Sexual imprinting can induce sexual preferences for exaggerated parental traits. *Current Biology* 16: 1128–1132.

ten Cate, C., and D. R. Vos, 1999. Sexual imprinting and evolutionary processes in birds: a reassessment. *Adv. Stud. Behav.* 28: 1–31.

ten Cate, C., D. R. Vos, and N. Mann. 1993. Sexual imprinting and song learning: two of one kind? *Neth. J. Zool.* 43: 34–45.

Thielcke, G. 1973. On the origin of divergence of learned signals (songs) in isolated populations. *Ibis* 115: 511–516.

Tonnis, B., P. R. Grant, B. R. Grant, and K. Petren. 2004. Habitat selection and ecological speciation in Galápagos warbler finches (*Certhidea olivacea* and *C. fusca*). *Proc. R. Soc. B* 272: 819–826.

Trauth, M. H., M. A. Maslin, A. Deino, and M. R. Strecker. 2005. Late Cenozoic moisture history of East Africa. *Science* 309: 2051–2053.

Travisiano, M., and P. B. Rainey. 2000. Studies of adaptive radiation using model microbial systems. *Amer. Nat.* 156 supplement: S35–S44.

Tudhope, A. W., C. P. Chilcott, M. T. McCulloch, E. R. Cook, J. Chappell, R. M. Ellam, D. W. Lea, J. M. Lough, and G. B. Shimmield. 2001. Variability in the El Niño-Southern oscilllation through a glacial-interglacial cycle. *Science* 291: 1511–1517.

Turelli, M., and H. A. Orr. 1995. The dominance theory of Haldane's rule. *Genetics* 140: 389–402.

Turney, C.S.M., A. P. Kershaw, S. C. Clemens, N. Branch, P. T. Moss, and L. K. Fifield. 2005. Millenial and orbital variations of El Niño/Southern Oscillation and high-latitude climate in the last glacial period. *Nature* 428: 306–310.

Valentine, J. W. 1973. *Evolutionary paleoecology of the marine biosphere.* Prentice-Hall, Englewood Cliffs, NJ.

Valentine, J. W. 2004. *On the origin of phyla.* University of Chicago Press, Chicago, IL.

Valentine, J. W., ed. 1985. *Phanerozoic diversity patterns.* Princeton University Press, Princeton, NJ.

Van Doorn, G. S., U. Dieckmann, and F. J. Weissing. 2004. Sympatric speciation by sexual selection: a critical reevaluation. *Amer. Nat.* 163: 709–725.

van Riper III, C., S. G. van Riper, M. L. Goff, and M. Laird. 1986. The epizootiology and ecological significance of malaria in Hawaiian birds. *Ecol. Monogr.* 56: 327–344.

Van Tuinen, M., and S. B. Hedges. 2001. Calibration of avian molecular clocks. *Mol. Biol. Evol.* 18: 206–213.

Veen, T., T. Borge, S. C. Griffiths, G.-P. Sætre, S. Bures, L. Gustafsson, and B. C. Sheldon. 2001. Hybridization and adaptive mate choice in flycatchers. *Nature* 411: 45–50.

Vincek, V., C. O'hUigin, Y. Satta, N. Takahata, P. T. Boag, P. R. Grant, B. R. Grant, and J. Klein. 1996. How large was the founding population of Darwin's finches? *Proc. R. Soc. B* 264: 111–118.

Vitt, L. J., and E. R. Pianka. 2005. Deep history impacts present-day ecology and biodiversity. *Proc. Natl. Acad. Sci. USA* 102: 7877–7881.

Vollmer, S. V., and S. R. Palumbi. 2002. Hybridization and the evolution of reef coral diversity, *Science* 296: 2023–2025.

Vrba, E. S., G. H. Denton, T. C. Partridge, and L. H. Burckle. 1995. *Paleoclimate and evolution, with emphasis on human origins.* Yale University Press, New Haven, CT.

Waddington, C. H. 1953. Genetic assimilation of an acquired character. *Evolution* 7: 118–126.

Wake, D. B. 2006. Problems with species: patterns and processes of species formation in salamanders. *Ann. Missouri Bot. Gard.* 93: 8–23.

Wallace, A. R. 1855. On the law which has regulated the introduction of new species. *Annals of the Magazine of Natural History 2nd series*, 16: 184–196.

Wallace, A. R. 1871. *Contributions to the Theory of Natural Selection. A series of essays.* Macmillan & Co., London, U.K.

Wara, M. W., A. C. Ravelo, and M. L. Delaney 2005. Permanent El Niño-like conditions during the Pliocene warm period. *Science* 2005: 758–761.

Weiblen, G. D. 2002. How to be a fig wasp. *Annu. Rev. Entomol.* 47: 299–330.

Weir, J. T. 2006. Divergent timing and patterns of species accumulation in lowland and highland neotropical birds. *Evolution* 60: 842–855.

Werner, T. K., and T. W. Sherry. 1987. Behavioral feeding specialization in *Pinaroloxias inornata*, the "Darwin's Finch" of Cocos Island, Costa Rica. *Proc. Natl. Acad. Sci. USA* 84: 5506–5510.

West Eberhard, M. J. 2003. *Developmental plasticity and evolution.* Oxford University Press, New York, NY.

White, W. M., A. R. McBirney, and R. A. Duncan. 1993. Petrology and geochemistry of the Galápagos Islands: portrait of a pathological mantleplume. *J. Geophys. Res.* 98: 19533–19563.

Whiteman, N. K., S. J. Goodman, B. J. Sinclair, T. Walsh, A. A. Cunningham, L. D. Kramer, and P. G. Parker. 2005. Establishment of the avian disease vector *Culex quinquifasciatus* Say, 1823 (Diptera: Culicidae) on the Galápagos Islands, Ecuador. *Ibis* 147: 844–847.

Wiggins, I. L. 1966. Origins and relationships of the flora of the Galápagos Islands. In R. I. Bowman, ed., *The Galápagos*, 175–182. University of California Press, Berkeley, CA.

Williams, E. E. 1969. The ecology of colonization as seen in the zoogeography of anoline lizards on small islands. *Quart. Rev. Biol.* 44: 345–389.

Williams, E. E. 1972. The origin of faunas: evolution of lizard congeners in a complex island fauna—a trial analysis. *Evol. Biol.* 6: 47–89.

Willis, B. L., M.J.H. van Oppen, D. J. Miller, S. V. Vollmer, and D. J. Ayre. 2006. The role of hybridization in the evolution of reef corals. *Annu. Rev. Ecol. Evol. Syst.* 37: 489–517.

Wilmé, L., S. M. Goodman, and J. U. Ganzhorn. 2006. Biogeographic evolution of Madagascar's microendemic biota. *Science* 312: 1063–1067.

Wilson, E. O. 1992. *The diversity of life.* Harvard University Press, Cambridge, MA.

Woodward, F. I. 1987. *Climate and plant distribution.* Cambridge University Press, Cambridge, U.K.

Wright, J. W. 1983. The evolution and biogeography of the lizards of the Galápagos Archipelago: evolutionary genetics of *Phyllodactylus* and *Tropidurus* populations. In R. I. Bowman, M. Berson, and A. E. Leviton, eds., *Patterns of evolution in Galápagos organisms*, 123–155. American Association for the Advancement of Science, Pacific Division, San Francisco, CA.

Wright, S. 1932. The role of mutation, inbreeding, crossbreeding, and selection in evolution. *Proc. 6th Internat. Congr. Genet.* 1: 356–366.

Wright, S. 1940. The statistical consequences of Mendelian heredity in relation to speciation. In J. S. Huxley, ed., *The new systematics*, 161–183. Oxford University Press, Oxford, U.K.

Wright, S. 1977. *Evolution and the genetics of populations.* Vol. III, *Experimental results and evolutionary deductions.* University of Chicago Press, Chicago, IL.

Wu, C.-I., and A. W. Davis. 1993. Evolution of post-mating reproductive isolation—the composite nature of Haldane's rule and its genetic basis. *Amer. Nat.* 142: 187–212.

Wu, P., T.-X. Jiang, S. Suksaweang, R. B. Widelitz, and C.-M. Chuong. 2004. Molecular shaping of the beak: a paradigm for multiple primordial morphogenesis. *Science* 305: 1465–1467.

Yamagishi, S., and M. Honda. 2005. Tracking the route taken by Rufous Vangas. In S. Yamagishi, ed., *Social organization of the Rufous Vanga: the ecology of Vangas—birds endemic to Madagascar*, 141–162. Kyoto University Press, Kyoto, Japan.

Yamagishi, S., and K. Eguchi. 1996. Comparative foraging ecology of Madagascar vangids (Vangidae). *Ibis* 138: 283–290.

Yang, S. Y., and J. L. Patton. 1981. Genic variability and differentiation in Galápagos finches. *Auk* 98: 230–242.

Zachos, J., M. Pagani, L. Sloan, E. Thomas, and K. Billups. 2001. Trends, rhythms, and aberrations in global climate 65 Ma to present. *Science* 292: 686–693.

Zhang, R. H., L. M. Rothstein, and A. J. Busalacchi. 1998. Origins of upper-ocean warming and El Niño changes on decadal scales in the tropical Pacific Ocean. *Nature* 391: 879–883.

Zimmerman, E. C. 1948. Introduction. *Insects of Hawaii*, vol. 1. University of Hawaii Press, Honolulu, HI.

Zink, R. M. 2002. A new perspective on the evolutionary history of Darwin's finches. *Auk* 119: 864–871.

Zink, R. M., and M. C. McKitrick. 1995. The debate over species concepts. *Auk* 113: 701–719.

Author Index

206

Subject Index

211